Unlimited Impossibilities:

Intelligence Support to the *Deepwater Horizon* Response

by CAPT Erich M. Telfer, USCG

N|PRESS
NATIONAL INTELLIGENCE UNIVERSITY

National Intelligence University
Washington, DC

January 2014

In *Unlimited Impossibilities: Intelligence Support to the Deepwater Horizon Response*, CAPT Erich M. Telfer of the U. S. Coast Guard examines the Intelligence Community and Coast Guard response to the 2010 BP oil explosion on *Deepwater Horizon*, which spilled tons of oil into the Gulf of Mexico. Telfer points out problems that hampered the response and suggests approaches for better communication and coordination in future responses.

The goal of the NI Press is to publish high quality, valuable, and timely books on topics of concern to the Intelligence Community and, more broadly, the U.S. government. Books published by the NI Press undergo peer review by senior officials in the U.S. government and outside experts.

How to order this book. Everyone may download a free electronic copy from our website at *http://www.NI-U.edu* U.S. government employees may request a complimentary copy by contacting us at: The general public may purchase a copy from the Government Printing Office (GPO) at *http://bookstore.gpo.gov.*

Editor, NI Press
Office of Research
National Intelligence University
Defense Intelligence Agency
Joint Base Anacostia-Bolling
Washington, DC 20340-5100

ABSTRACT

In April 2010, when BP's DEEPWATER HORIZON rig exploded and spilled millions of barrels of oil in the Gulf of Mexico, the United States had no national-level plan for intelligence support to guide the process of conducting intelligence collection, analysis, production, and dissemination. The intelligence effort lacked unity of command, and took over a month and a half to develop a functioning system that served strategic- and operational-level decisionmakers well. Even then, it was of less help to tactical responders. A considerable amount of data was collected during the response (primarily by remote sensing), but analyzing, producing, and disseminating the subsequent intelligence proved difficult. The lack of an air tasking order directing and deconflicting *all* flights above and around the spill resulted in near air-to-air collisions and hampered intelligence collection. In addition, no imagery-based common operating picture existed to display the spill, the impacted areas, and the response effort.

To prevent these problems from happening again, the Coast Guard should develop a plan for future spills of national significance that describes the organization, command relationship, function, and goal of intelligence support. The Coast Guard also should establish a system to manipulate and share imagery of the spill among federal, state, local, tribal, private, and public organizations. An air tasking order should guide obtaining this imagery, to ensure flight safety and intelligence collection. Finally, a system for developing a common operating picture should be used to share information and manage intelligence across the spectrum of responders and response activities.

ACKNOWLEDGMENTS

It would be a fool's errand to attempt a work such as this without the dedicated assistance of many people. I would like to acknowledge them. First, I wish to thank Dr. Cathryn Thurston, who was my faculty chair at the Center for Strategic Intelligence Research at the National Intelligence University. Dr. Thurston served as a guide, mentor, editor, and champion throughout my many months of research, writing, and editing. This project is exponentially better because of her steady insights and suggestions (most of which I heeded). However, I heeded all the suggestions of my editor, "George," and his team, whose skill and patience uncovered errors great and small, and turned this from a project into a worthwhile endeavor.

Thank you to those who participated in the interviews. These professionals gave their time, support, and perspectives freely. In some cases, they spent hours explaining their experiences and observations about intelligence and the *Deepwater Horizon* response. Without their cooperation, my work would have been impossible.

To CDR Kim Nettles, USCG (ret.), I offer grateful appreciation. A veteran of reviewing my writing, CDR Nettles encouraged this topic and then bravely agreed to read and edit my project in anticipation of publication. CDR Nettles also provided the strategic vision to help me better understand my audience and the impact of this work.

A special thanks goes to my good friend LTC Mike Trevett, USA, author of the book *Isolating the Guerrilla*. Mike's 20-plus years of leadership and encouragement have motivated me not only to look at what is not working properly, but to try and fix it as well.

Most importantly, I thank my wife, Jenn, and our children, Lauren, Daniel, and Savannah. They have had the courage and love to steadfastly follow me from coast to coast, and from the U.S. heartland to Mexico City. I couldn't be more proud of them and grateful for their support, love, and good humor during it all.

PREFACE

The story of the *Deepwater Horizon* actually begins on March 24, 1989, when the oil tanker EXXON VALDEZ struck the Blight Reef in Prince William Sound, Alaska, spilling up to 750,000 barrels of oil.[1] This was the most significant maritime oilspill since the 1969 blowout of the Union Oil Company platform in the Santa Barbara channel off the coast of California. As had become the practice, the federal government and the oil industry worked together to review safety procedures and practices used on the EXXON VALDEZ.[2] From the *Valdez* collision and spill came the Oil Pollution Act of 1990.

In addition to the Oil Pollution Act of 1990, the U.S. Congress gave the U.S. Coast Guard additional funding to increase the size of the officer corps to be better staffed and prepared to respond to future spills of national significance (SONS).[3] The Coast Guard added an entire Officer Candidate School class in 1991 using this additional Congressional funding. (I know this because I was among those 59 newly commissioned officers.)

On April 20, 2010, an explosion erupted on the deep-sea drilling rig DEEPWATER HORIZON that required the largest oilspill response in the history of the nation.[4] The spill was the first environmental disaster officially designated as a spill of national significance. In addition to the tens of thousands of people who responded to the spill from dozens of agencies, intelligence officers from several organizations deployed to the Gulf Coast in order to support the response effort. But out of all the coverage of the explosion and spill, few, if any, of the thousands of media pieces even mention the intelligence support effort. The official reports are nearly silent on the topic, and the books published in early 2011 do not mention the intelligence support to the spill response. Judging by the reporting, it is almost as though there was no intelligence support to the *Deepwater Horizon* response. But intelligence officers and their parent agencies *did* respond to the spill by supporting strategic, operational, and tactical decisionmakers in battling the spill. They coordinated intelligence among multiple federal and state agencies and departments. They managed and guided multiple satellite and aircraft systems providing hundreds of images of the spill. Unfortunately, nothing has been written about intelligence support to the *Deepwater Horizon* response.

This book documents the Intelligence Community response to the spill, and assesses the successes and failures of its efforts.

Contents

Acknowledgments v

Preface vii

Chapter 1—Intelligence and Disaster Response 1

To the "Right of Boom": U.S. Government Response to
Deepwater Horizon 1

Defining Intelligence for Disaster Response 3

The Role of Intelligence 6

Intelligence Support to Disaster Response—Past and
Current Plans 10

Graduate Papers 19

Much Has Been Written about *Deepwater Horizon*, Just Not about
Intelligence 21

Books—Recent and Detailed (Except about Intelligence) 29

Summary 32

Chapter 2—Methodology: Authentic Answers to Authentic
Questions 35

Conduct of the Interviews 38

Chapter 3—Insufficient Intelligence Plans Hampered Response 43

Lack of Imagination Hampered Planning and Preparation 43

Lack of Command and Control 45

Geospatial Information Systems (GIS) and Human Intelligence
Support to Deepwater Horizon 48

Reactive Intelligence Slowed the Response 50

Findings and Recommendations 52

Chapter 4—The Deepwater Horizon Intelligence Cycle:
Spinning But Off Balance 55

Contents (continue)

Step 1: Establishing Intelligence Requirements 55

Step 2: Intelligence (Dis)organization: Collection Planning 67

Step 3: Collection 68

Step 4: Analysis 89

Findings and Recommendations 95

Chapter 5—The Common Operating Picture 97

Production and Dissemination 97

Findings and Recommendations 109

Chapter 6—Conclusion 111

Chapter 7—Epilogue 115

Lessons Experienced Are Not Lessons Learned 115

Appendix—SONS Priority Intelligence Requirements 121

Findings and Recommendations 123

Notes 127

Bibliography 145

About The Author 151

Chapter 1
Intelligence and Disaster Response

> "The contemporary has no perspective; everything is in the foreground and appears the same size. Little matters loom big, and great matters are sometimes missed because their outlines cannot be seen."[5]

—Barbara Tuchman

To the "Right of Boom": U.S. Government Response to *Deepwater Horizon*

The crew of the oil rig DEEPWATER HORIZON was conducting exploratory drilling on the Macondo 252 well on April 20, 2010, when the rig exploded, killing 11 men and releasing an estimated 53,000 barrels of oil per day into the Gulf of Mexico for 87 days.[6] After the Coast Guard completed the initial search and rescue, the U.S. government launched an environmental response unparalleled in U.S. history. Department of Homeland Security (DHS) Secretary Janet Napolitano declared the situation to be a spill of national significance on April 29. It was the first spill to officially carry this designation, and was the largest spill since the *Exxon Valdez* disaster.[7] On the same day, Secretary Napolitano directed Admiral Thad Allen, U.S. Coast Guard commandant, to be the National Incident Commander (NIC), and the Coast Guard to be the lead agency. As Rear Admiral Peter Neffenger detailed in his September 22, 2010, testimony before the House Committee on Homeland Security, "At its peak, we deployed more than 47,000 responders, over 3,000 of which were Coast Guard members; 4 million feet of boom; more than 7,000 vessels, including 835 specialized skimmers; over 3,000 vessels of opportunity; 120 aircraft; and hundreds of public and private organizations and volunteers."[8] The Coast Guard alone mobilized 14 percent of its total workforce, active duty and reserves, to respond to the spill.[9] Admiral Allen stayed on as the NIC until October 2010, even after Admiral Robert Papp took over as Coast Guard commandant on May 25. Admiral Allen has said that the response effort he led should not be thought of as an isolated incident, commenting to Eugene Robinson of the *Washington Post* that "it would be adding a crime to a crime if we didn't make this one of the great

CAPT Erich M. Telfer

learning laboratories in the history of this country."[10] In that spirit, this book reviews what has been learned about the *Deepwater Horizon* incident so far and, more specifically, what role, and to what degree of success, intelligence played in supporting the response effort.

> *Decisionmakers at the strategic, operational, and tactical levels needed timely intelligence from the moment the Macondo well exploded until after it was capped in July.*

Decisionmakers at the strategic, operational, and tactical levels needed timely intelligence from the moment the Macondo well exploded until after it was capped in July. They needed to know the location of the oil, where the oil was going, where their response personnel and assets were, and what efforts were being made to fight the spill.

Intelligence and Disaster Response: A Review of Literature

Intelligence has a role in disaster response, and it definitely played a role in *Deepwater Horizon*, yet a review of U.S. government plans, "lessons learned" documents, academic work, media coverage, and books shows little attention has been paid to this topic. National disasters of the magnitude of Hurricane Katrina or *Deepwater Horizon* share the attributes of being unexpected, chaotic, and requiring national-level assistance to the local response. When the federal, state, and local responders arrive at a disaster and begin their work, the most pressing and important requirement is that of managing information and providing intelligence to decisionmakers at all levels.

Intelligence enables decisionmakers to apply resources and people in the most efficient, timely manner to save lives and property. In the area of disaster response, though, the term "intelligence" is not well understood, even though many U.S. government publications discuss disaster response and intelligence to varying degrees. There are a few graduate papers that speak directly to the role of intelligence in disaster response based on the experience of Hurricane Katrina. Several federal agencies deployed intelligence officers to the *Deepwater Horizon* response to assist strategic, operational, and tactical decisionmakers. These officers used several intelligence functions in their work and Rear Admiral Neffenger used the word "superb" when remarking on their effort.[11] Presidential and U.S. government "lessons learned" reports on the *Deepwater Horizon* response barely touch on the issue of intelligence support. And media reporting and recently published books speak sparingly, with a few notable exceptions, about the contribution of intelligence to managing the disaster.

Defining Intelligence for Disaster Response

The term "intelligence" is not well defined in relation to disaster response and so a description will aid in focusing the discussion. This section broadly describes *intelligence* and then, more specifically, delineates intelligence from *operational information* used during tactical actions.

First of all, the word "intelligence" carries a weighty, broad, and ill-defined connotation not at all appropriate when talking about natural disasters. Television shows and movies abound with the derring-do of "intelligence officers" who often behave as well-armed, reckless saboteurs who do little or nothing resembling actual intelligence work.[12] "Intelligence," too, is often confused with "information." A good example of this confusion is shown in the *National Geographic Explorer* documentary *Can the Gulf Survive?*, when the narrator said that "intelligence" from a command post directed the U.S. Coast Guard cutter *Elm* to an oil slick.[13] What the documentary should have said was simply that the command post directed the cutter to the oil. "Intelligence" does not *direct* operations, although a commander may use intelligence to those ends. This may appear a fine distinction, but it is an important one to the Intelligence Community. A definition of "intelligence" for the scope of this project is therefore warranted.

Coast Guard Publication 2-0, *Intelligence,* puts it this way, "Intelligence is the development and analysis of raw material in order to determine what the information means and to identify the implications for decisionmaking."[14] "In other words," Pub 2-0 goes on to say, "intelligence is the analysis and synthesis of information into knowledge."[15] *Information* is material that has not been evaluated. "The purpose of intelligence, therefore," the Coast Guard publication says, "is to inform commanders and decisionmakers by providing accurate, timely, and relevant knowledge about adversaries, threats, and the surrounding environment."[16] Based on my professional experience in Coast Guard intelligence and graduate study on the topic, I suggest a further refinement when discussing intelligence, even though the definitions from Coast Guard Publication 2-0 are perfectly fine. My refinement deals with the collection and analysis of information and the purpose of intelligence within disaster response.

CAPT Erich M. Telfer

Information must be *collected* (i.e., gotten from somewhere) before a person may analyze it. This means someone must develop and implement a plan to collect the information. It has been my experience that the term "analysis" in the definition of intelligence is often over-thought. *Analysis* means reflection and perhaps study on the collected information based on one's experience and understanding. A commanding officer of a Coast Guard cutter may analyze a piece of information during a tactical situation and use his judgment to make a decision based on that analysis. If the outcome of the analysis is of greater breadth and depth than the initial information, then the commanding officer has turned the information into *intelligence*. Coast Guard cutter commanding officers did this during the *Deepwater Horizon* response when Air National Guard aircraft passed live video feeds directly to the cutter. Another example would be the integration of the Environmental Response Management Application (ERMA) within the various command posts in July 2010 into a common operating picture (COP). Raw images and data points about oil and weather information are not intelligence. However, when ERMA pulled that data (and a great deal more), placed it on a geospatial information system, and made it available to federal, state, local, and private company decisionmakers who then took the data ERMA had contextualized, that moved closer to a more traditional understanding of intelligence.

A Note on the Difference between Operations and Intelligence

The collection and analysis of intelligence should be understood as different from *operations*, which include observing, orienting, deciding, and acting (also known as the acronym "OODA Loop"), as shown in Figure 1. The U.S. Air Force developed the OODA Loop after the Korean War to describe the process a fighter pilot uses during operations. Other communities within the military services adopted the OODA Loop description to explain the *operational* information process. The Coast Guard teaches and practices a variation of the OODA Loop consisting of the steps surveil, detect, classify, identify, and prosecute (SDCIP), shown in Figure 2. The SDCIP does essentially the same thing as the OODA Loop in terms of describing a process of prosecuting a mission or action. Both methods begin with observation and end with action. They are loops because the commander continually reassesses the situation and makes decisions based on his observations. But these are "operational information" methods, not intelligence.

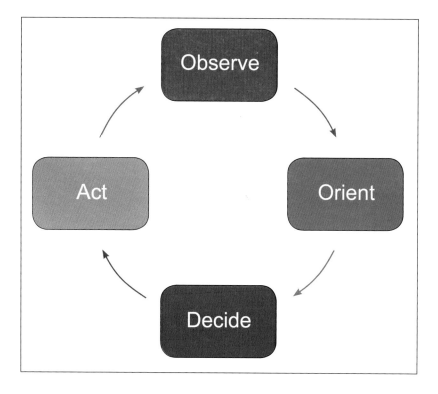

Figure 1: An operational information process: The "OODA Loop".

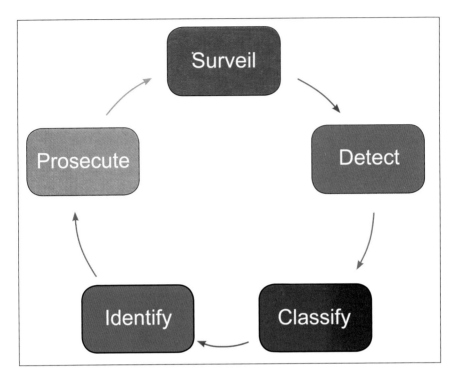

Figure 2: The Coast Guard version: The "SDCIP" Loop.

The Role of Intelligence

Intelligence has two purposes: to warn decisionmakers, and to aid decision-makers in making better decisions. That is all. The intelligence cycle is complete when the decisionmaker actually makes a decision, even if that decision is to do nothing. The great parade of intelligence agencies that manage the nation's collection capabilities and personnel largely guide themselves based on those two purposes. It is therefore important to understand that decisionmakers, whether at the local, state, or national level, drive the intelligence cycle. They do this

> *Intelligence has two purposes: to warn decisionmakers, and to aid decisionmakers in making better decisions. That is all.*

by telling the intelligence officers assigned to them what the decisionmaker needs to know. These are called *intelligence requirements.*

Intelligence also has a kind of OODA Loop all its own. In this case, the following response functions adhere to what is called the *intelligence cycle*: requirements, planning, collection, analysis, production, and dissemination. First, a decisionmaker has a need for expanded knowledge to prosecute a response (requirement), and then the intelligence officer develops a way to get that information (organization).[17] Next, an asset is directed to obtain the information (collection) and the collected information is studied for meaning (analysis).[18] Lastly, the intelligence is put in a format useable for the decision-maker (produced) and presented to the decisionmaker in a timely understood medium (dissemination).[19] Figure 3 shows this classic, basic intelligence cycle, which is generally depicted in a circle. The intelligence cycle is immutable whether searching for Iranian nuclear sites or oil gushing up from the ocean floor. It is an ordered, iterative process used to focus the intelligence staff on providing the decisionmaker with what he or she needs to warn and assist in making a decision.

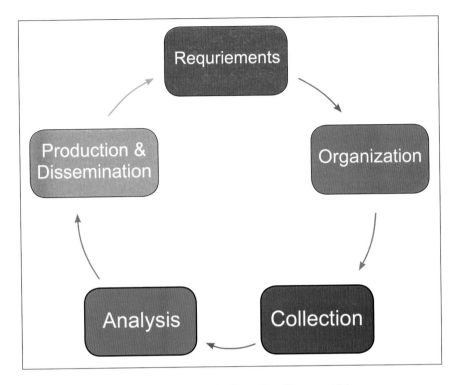

Figure 3: The basic intelligence cycle, from Intelligence 101.

An Argument for Applying the Intelligence Cycle to the *Deepwater Horizon* Response

I include those *Deepwater Horizon* response personnel and certain functions that meet the preceding description under the "intelligence" umbrella: remote sensing, geospatial information systems, satellite imagery collection, airborne imagery collection, and shoreline cleanup assessment teams (SCATs).

I understand that this is an expanded view of intelligence within the context of environmental response and a spill of national significance. It is true that those Coast Guard members who work in environmental response generally do not use the term *intelligence* in this context. Most Coast Guard members whose careers center on environmental protection and oilspill response (the

"M" community, as the Coast Guard still refers to it) generally do not practice the intelligence cycle in their normal operations.

Perhaps, in routine response operations, there is not a pressing need to define intelligence support so specifically. However, at the national response level during declared disasters and spills of national significance, the Coast Guard response community has an insufficient understanding of intelligence support—or even the need for it. As an illustration, a senior Coast Guard captain who both spent a career in the maritime (environmental) response community and responded to the *Deepwater Horizon* spill commented to me that intelligence had no place in environmental response.[20]

This study intends to explain the role traditional intelligence practice played in the *Deepwater Horizon* response. Although few environmental response operations may necessitate *intelligence* support, the *Deepwater Horizon* spill clearly illustrates that a spill of national significance does require a robust intelligence presence that is properly organized.

The Role of Intelligence in a Federal Disaster Response Context

A good place to begin the discussion is with a review of U.S. government documents that guide disaster response and try to outline the role of intelligence within that context. The U.S. government struggles with defining the role of intelligence in disaster response at the federal level. This struggle is reflected in the various plans and frameworks that were developed for disaster response where the importance of *intelligence* is described in various levels of detail. In short, the plans and frameworks outline some intelligence responsibilities, and encourage intelligence and information sharing. But the documents do not establish clear unity of command in describing where the intelligence function should reside in disaster response. The guides give too much latitude to senior decisionmakers regarding where to place the intelligence function, especially considering that these decisionmakers generally have little to no understanding of the capabilities or the role of intelligence. The guides neither establish a principal intelligence officer within the response construct, nor do they distinguish if the intelligence function should operate differently for

> *... documents do not establish clear unity of command in describing where the intelligence function should reside ...*

a spill of national significance than for other responses. There has been some thoughtful work done by graduate students to address these issues, but considering the substantial role intelligence support played in the response to *Deepwater Horizon*, the topic has been grossly understudied.[21]

Intelligence Support to Disaster Response—Past and Current Plans

Going back as far as the early 1990s, the Federal Response Plan discussed intelligence, as did the subsequent National Response Plan. The current National Response Framework also touches briefly on the role of intelligence. Two other current federal documents, the *National Incident Management System* and the *National Infrastructure Protection Plan*, also mention intelligence as it relates to disaster response.

The Stafford Act and the Federal Response Plan of 1992

The first of these models, the *Federal Response Plan* (FRP), incorporated the efforts of 27 federal agencies to implement the Stafford Act, which included intelligence support to law enforcement and other agencies during a disaster response. The Robert T. Stafford Disaster Relief and Emergency Assistance Act[22] authorized most federal disaster response activities under the U.S. Federal Emergency and Management Agency (FEMA), the lead agency for federal disaster response. The Stafford Act defines a major disaster as "any natural catastrophe . . . or, regardless of cause, any fire, flood, or explosion, in any part of the United States" that the President determines requires federal assistance to the states to alleviate damage, loss, and suffering.[23] The FRP is federal support to state and local response efforts where they have the lead and the federal government provided assistance. The FRP is the methodology the federal government uses when the President declares a major disaster.

Intelligence in the FRP

Within the FRP, intelligence is described in law enforcement terms and for situational awareness. Regarding intelligence and law enforcement, intelligence may be used by the federal government to assist state governments that face a law enforcement emergency during which the collection and preservation of evidence is required for further investigation and prosecution.[24] For example, the response to a bomb detonation would include evidence gathering, referred to as "intelligence" in parts of the FRP, as law enforcement authorities search for the people responsible. Regarding situational awareness, the plan lists intelligence as a support function of response planning and daily briefings.[25]

The FRP defines intelligence functions as *components* of disaster response, which is more important for our purpose here, including remote sensing and situational awareness. Embedded in the FRP are also the Emergency Support Functions, which are annexes that delineate and organize the type of help the federal government may provide to the states. Emergency Support Function 5, the Information and Planning Annex, describes the role and organization of intelligence regarding remote sensing. Within the Information and Planning Section resides the Technical Service Branch, which includes a remote-sensing specialist, geospatial information systems (GIS) coordinator, GIS specialist, and technical specialists. The role of the Technical Service Branch includes coordinating remote-sensing reconnaissance requirements, establishing and maintaining a geographic information system, providing hazard-specific technical advice to operational planning, and using other experts (e.g., meteorologists) as required.

The FRP also addresses the role of intelligence and situational awareness. The FRP includes assumptions for the Planning Section such as the "immediate and continuous demand by officials" for information that provides operational and strategic decisionmakers with information and intelligence. This demand for continual information by senior leaders, described by Admiral Allen as "insatiable," was much in evidence during *Deepwater Horizon*. Also listed in the Information and Planning Appendix is a potential need to "rapidly deploy field observers" to collect information.

National Oil and Hazardous Substance Pollution Contingency Plan

The National Oil and Hazardous Substance Pollution Contingency Plan (40 CFR Part 300), or National Contingency Plan (NCP), was established in 1968 as the U.S. blueprint for responding to both oilspills and hazardous substance releases.[27] The NCP authorizes the authorities of the federal government to organize for and respond to maritime oil and hazardous material spills. The main tenets of the NCP are that the On Scene Coordinator directs all federal, state, and private response activities. Congress has modified the NCP since 1968, most recently in 1994 to encompass the

Oil Pollution Act of 1990. The NCP directs that the federal government have the lead in responding to a spill of national significance. This is different from Stafford Act federal responses activity where the states take the lead role and the federal government supports them. The *Deepwater Horizon* response was managed under the NCP with the federal government in the lead. This was to cause confusion among the impacted state governments, all of whom were accustomed to the practices of the Stafford Act. While the NCP does not discuss intelligence specifically, it does promote unity of command and unity of effort in the response operations.

The National Response Plan of 2003

In February 2003, President George W. Bush issued a Homeland Security Directive (HSPD-5) ordering the secretary of the Department of Homeland Security to draft a National Response Plan.[28] The *National Response Plan* (NRP) superseded the *Federal Response Plan* and was enacted "to align Federal coordination structures, capabilities, and resources into a unified, all-discipline, and all-hazards approach to domestic incident management. This approach is unique and far-reaching in that it, for the first time, eliminates critical seams and ties together a complete spectrum of incident management activities to include the prevention of, preparedness for, response to, and recovery from terrorism, major natural disasters, and other major emergencies."[29]

Intelligence in the NRP

Included in the section on Incident Management Activities is the idea that incidents of national significance may have results "far beyond" the immediate area; then, the NRP provides a framework to manage the additional impacts.[30] But the only intelligence function discussed in this section is counterintelligence (also known as counterespionage), which is included among a list of possible countermeasures that may be required after a disaster to safeguard personnel and classified material. Intelligence is also discussed as a role authorized under the attorney general when used "to detect, prevent, preempt, and disrupt terrorist attacks against the United States."[31]

The Homeland Security Operations Center

The NRP's description of the Homeland Security Operations Center (HSOC) is much more useful. The HSOC is the hub for operational coordination and strategic awareness in response to a domestic incident. Specifically, the Intelligence/Information Analysis subsection describes DHS as "responsible for interagency intelligence collection requirements, analysis, production, and production dissemination" (the *intelligence cycle*).[32] The plan outlines the intelligence tasks for DHS including coordinating and disseminating threat warnings, coordinating with other federal agencies regarding counterterrorism, providing analytical support, providing threat awareness, and maintaining "real-time" communication with other intelligence organizations. The NRP also lists and describes several other federal organizations that conduct intelligence as part of homeland security, including the FBI's Strategic Information and Operations Center and the National Counter Terrorism Center.

The Joint Field Office and Intelligence Support

The Joint Field Office is an organization used to support on scene responders. The NRP explains that the Joint Field Office Coordination Group would determine the placement of the intelligence function based on the incident and the situation. The NRP says that the intelligence function at the operational level may be placed within the planning or operations sections of a joint field office (JFO) or as a standalone section. The JFO sections include operations, planning, logistics, and finance and administration (comptroller), and, if needed, a sixth section, intelligence. However, the common practice during a response is to place intelligence within the staff sections of either planning or, less frequently, operations. Just like the *Federal Response Plan*, the NRP includes a geospatial information systems function within the Planning Section for Emergency Support Function #5, the Emergency Management Appendix.

The Failure of the NRP When Used during Hurricane Katrina

Although the *National Response Plan* expanded the role of intelligence when compared with the *Federal Response Plan*, the NRP still did not fully meet U.S. government needs. The plan failed the test of Hurricane Katrina because it went neither far enough nor broadly enough in detailing what was really needed in response to a disaster of this magnitude. According to James J. Carafono and Richard Weitz's critique of the Katrina response, *Mismanaging Mayhem: How Washington Responds to Crisis*, the NRP was not a doctrine, strategy, system, or plan as commonly understood.[33] It was, instead, a practice. Traditional planning consists of five parts: mission statement, situation description, tasks for subordinate elements, logistics and administration, and command and communications. The NRP lacked these elements as well as national-level direction on intelligence support to the response effort. Intelligence was directed and described on a basic level, but it appears the organizations contributing to the response, the individual agencies, were left to their own understanding and practice of intelligence support. In other words, the NRP listed the intelligence ingredients, but not how to mix and prepare them for success. Consequently, the NRP was scrapped in favor of the National Response Framework of 2008.

National Reflection on the Role of Intelligence: The National Response Framework

The 9/11 attacks, combined with insufficient federal responses to Hurricanes Katrina and Rita, motivated significant reflection and study about how the government should address intelligence support to disaster response.[34] Congress intimately involved itself, observing, in 2003, that there was a lack of information and intelligence sharing among intelligence agencies and responders, and calling for formal, written intelligence-sharing agreements.[35] The culmination of this third iteration since the early 1990s was the *National Response Framework* (NRF).

The stated goal of the framework is to guide how the nation conducts all-hazard response. The NRF "explains the common discipline and structure" of response as opposed to being a response plan. The NRF is built on the *National Incident Management System* (described later).

Intelligence in the NRF

Within the NRF, intelligence is split into two general function areas: support to law enforcement for investigations and prosecutions, and the more traditional role of intelligence as support to national-level decisionmakers. The DHS secretary is designated as the federal official for domestic incident management. The NRF defines the Director of National Intelligence as the lead federal officer responsible for implementing the National Intelligence Program in the event of a response. The NRF goes on to list how important intelligence is in planning and how mechanisms must be established to share intelligence with local, tribal, and state actors. In describing the Incident Command System (ICS), upon which the National Incident Management System is built, intelligence is mentioned as a possible function area if deemed appropriate. This remains unchanged from the description in the *National Response Plan*. Further descriptions of intelligence in the *National Response Framework* mimic those in the rest of the *National Response Plan*.[36] With the NRF being a general guide, the U.S. government implements disaster response via the National Incident Management System.

Implementation of the NRF: The National Incident Management System

The *National Incident Management System* (NIMS) is the companion document to the *National Response Framework* and exists to "provide a consistent nationwide template" for federal, state, local, nongovernmental organization, and private-sector response to incidents with an "expanded intelligence function."[37] The NIMS describes intelligence as it relates to disaster response with greater detail than the *National Response Framework*. The cover letter by then-DHS Secretary Michael Chertoff explains that the latest iteration of the NIMS expands "the intelligence/investigations functions" of response. It does this by including intelligence sharing as a mitigating factor in reducing damage and loss of life, arguing for a standard incident reporting system to better share intelligence, and explaining the value of geospatial information systems in disaster response. NIMS outlines the value of GIS to a common operating picture, intelligence support to decisionmakers, and operational decisionmaking. The NIMS wisely cautions that the products of GIS require analysis by trained specialists since the misinterpretation of imagery may cause "inconspicuous [but] serious errors."[38] The NIMS contains 14 "management characteristics" that contribute to the strength and efficiency of the response, including information and intelligence management, which requires the organization to set up and use the intelligence cycle to assist during the incident response.

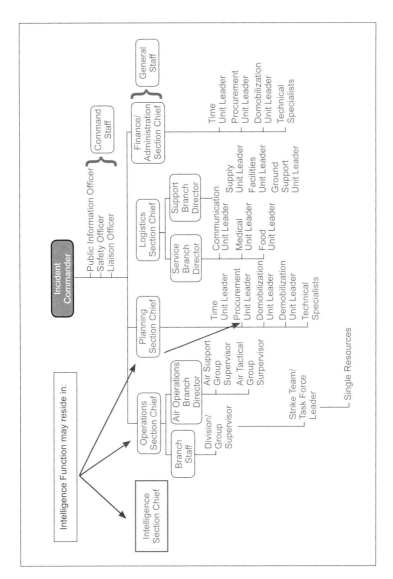

Figure 4: Organizational Terminology: ICS Organizational Chart

Source: U.S. Department of Agriculture, "ICS 200—Lesson 3: ICS Organization," file titled "Organizational Terminology: ICS Organizational Chart," accessed March 1, 2011, *http://www.usda.gov/documents/ICS200Lesson03.pdf.* Additions in red and "Intelligence Section Chief" box added by author.

The NIMS, therefore, helpfully characterizes the intelligence/investigation function in detail and where it resides in various sections of the Incident Command System. Intelligence as a function normally resides within the planning section, where it is responsible for the status of situations, resources, and anticipated incidents. Varying definitions of "intelligence" again lead to some misunderstanding, since the word "intelligence" is really referring to situational awareness, describing the events as they are and not traditional intelligence analysis. Situation awareness is a key factor in decisionmaking, but it is not *intelligence* as earlier discussed. However, when situational awareness is combined in a format that provides greater depth of understanding and adds meaning to the information (such as a GIS-based common operating picture should), then "intelligence" moves closer to the Intelligence Community's understanding of the term.

The NIMS goes on, however, to explain that the response organization must establish "a system for the collection, analysis, and sharing of information developed during intelligence/investigation efforts."[39] Now, the word *intelligence* is contributing to prosecuting criminal activities or in determining cause and impact of an incident. This level of intelligence support to the NIMS should be set up within the Incident Command System, "when there is a significant intelligence/investigations component to the incident."[40]

Confusion over Function and Role in the NRF

The dual role of intelligence support to disaster response as well as disaster investigations is an important point to keep in mind as this discussion moves forward, because the functions are quite different, they are not well defined, and it is unclear where their functions should lie within the National Response Framework. The response organization to *Deepwater Horizon* did not establish a separate intelligence/information function, nor was intelligence support ever well defined. The NIMS underscores that the intelligence/investigations function "has responsibilities that cross *all* interests of departments involved during an incident" even if certain functions remain specific to law enforcement.[41] (Emphasis added.) However, the NIMS discussion offers too much latitude in the organization of that function.

CAPT Erich M. Telfer

Implementing the Framework: The National Infrastructure Protection Plan

While the NIMS focuses on disaster response, the *National Infrastructure Protection Plan* (NIPP) aims to address deliberate manmade attacks against the country's critical infrastructure. The NIPP includes the use of intelligence as "one of the essential elements needed" to guard the country.[42] The NIPP is focused on the threat and response to terrorist attacks specifically (as opposed to natural disasters), but a spill of national significance could also be the goal or result of a terrorist attack.

Intelligence in the NIPP—More Robust Discussions

The NIPP includes access to "robust information-sharing networks that include relevant intelligence and threat analysis" as one of the four major goals toward which agencies must strive to protect critical infrastructure.[43] In fact, the NIPP mentions and addresses "intelligence" 126 times throughout the almost 200-page document. The NIPP argues that the new, post-9/11 terrorist threat requires "intelligence-driven analysis, information sharing, and unprecedented partnerships between the government and the private sector at all levels."[44] The NIPP describes an "all-hazards" approach to defining disasters and, specifically, lists natural and manmade events (including the *Exxon Valdez* oilspill) within the scope of the plan.

Intelligence sharing is a key theme in the NIPP. The NIPP highlights how multiple agencies at the federal, state, local, and tribal levels will respond to an attack or incident and why this necessitates information and intelligence sharing to ensure unity of effort. The NIPP goes so far as to stress a networked approach to bringing in state, local, tribal, territorial, and private-sector partners into the "intelligence cycle," including the development of intelligence requirements.[45] The NIPP discusses intelligence as it relates to evaluating "threat analysis and real-time incident reporting," such as a common operating picture.[46] The following discussion describes the many agencies and organizations that contribute some manner of intelligence function to protecting the nation's critical infrastructure. The discussion focuses on intelligence sharing, intelligence collaboration (including with geospatial information systems), and threat and risk analysis. Out of all the national disaster plans reviewed, the NIPP contains the most expansive and detailed approach to the role of intelligence in supporting disaster response.

Confusion in Function and Role in All Federal Plans

To summarize, the function and role of intelligence is not well defined within federal disaster response guides and plans. Intelligence sharing and coordination is promoted, but the plans establish neither unity of effort nor unity of command within the intelligence function. The plans do not discuss an intelligence methodology, the intelligence cycle, and give no direction as to *how* intelligence should support decisionmakers across the response effort. Incident leadership is given too much leeway in where to place the intelligence function and, historically, the leadership has most often placed the intelligence function within the Planning Section subordinate to the Situation Unit. This was the case during *Deepwater Horizon*. With the shortcomings of the U.S. government response plans in defining and guiding intelligence support to disaster response listed here, we will now turn to graduate-level academic work on the topic.

Graduate Papers

Two excellent papers about previous disaster response intelligence support stand out in the field. Major Jennifer Sovada's "Intelligence, Surveillance, and Reconnaissance Support to Humanitarian Relief Operations within the United States: Where Everyone Is in Charge" critically examined the need for intelligence, surveillance, and reconnaissance (ISR)[47] during domestic disaster relief.[48] Major Sovada looked at ISR in two case studies: the Hurricane Katrina response in September 2005 and the California wildfire response in October 2007. Major Sovada found that the efforts lacked unity of command and unity of effort. She is critical of the domestic Department of Defense (DoD) ISR effort, in addition to the interagency effort. Sovada argues that while DoD has improved domestic ISR since Hurricane Katrina and the California wildfires, many problems remain. Specifically, she writes that domestic ISR support to disaster relief requires an integrated, pre-established ISR plan.[49] Sovada also argues that ISR command relationships must be established before starting the response and that a robust liaison officer effort among the ISR entities will support unity of effort.[50] She says that legal issues regarding U.S. military domestic intelligence operations slowed and hampered ISR support during Hurricane Katrina and the wildfire responses, and that ISR must be included in domestic response exercises and practiced during those exercises.

Lieutenant Commander Joyce Dietrich, U.S. Coast Guard,[51] examined the relationship between intelligence dissemination and the emergency response after Hurricane Katrina in her thesis, "The Eyes of Katrina: A Case Study of Incident Command System (ICS) Intelligence Support during Hurricane Katrina." Dietrich found that the National Geospatial-Intelligence Agency (NGA), the National Oceanic and Atmospheric Administration (NOAA), the U.S. Air Force, the U.S. Coast Guard, and private companies provided intelligence products that "contributed to significant (tactical) successes during the Hurricane Katrina response."[52] Dietrich singles out NGA for specific praise, noting that the agency not only "was very creative in declassifying national asset imagery" for consumers, but that NGA was a large part of "what went right" during the intelligence support to the response.[53] She goes on to argue, however, that much of the imagery that NGA and the other organizations produced was not made available to the responders. Instead, the imagery was made available to strategic- and operational-level decisionmakers, but not those at the tactical level.[54] Dietrich highlights the role of private industry in supporting imagery, especially Google, which enabled hundreds of employees and private citizens to use its Google software to build imagery sets of New Orleans and overlay those sets on maps of the city. This assisted tremendously in assessing the storm's impact and damage.

But Dietrich observes that Admiral Thad Allen, the principal federal officer of the response, said that "communication between tactical responders and intelligence collectors was ad hoc" and that information was shared based on personal relationships, not on doctrine and planning. Successful intelligence sharing took place informally, crudely, and via liaison officers.[56] Intelligence was underutilized in part, Dietrich mentions, because the tactical responders had little knowledge and no training on the capabilities of national intelligence systems. Dietrich summarizes the key challenges her work uncovered like this:

> Despite . . . forward leaning actions of multiple intelligence agencies, many first responders never gained critical access to intelligence products that would have helped them save

additional lives. Communications between intelligence providers and field operators were scarce. A disproportionate number of senior individuals interviewed for this study identified a high level of interaction with the [Intelligence Community], while field level operators reported minimal to no open communication with intelligence components.

Dietrich could have been writing her conclusion about *Deepwater Horizon*, which again raises the question, what have we learned about intelligence support to national disasters?

Much Has Been Written about *Deepwater Horizon*, Just Not about Intelligence

A great deal has been written about the explosion, sinking, and subsequent oilspill of the *Deepwater Horizon*, but it rarely deals directly with intelligence support to the response. However, a review of what has been produced will frame the subsequent discussion. The coverage was voluminous, from the daily print and online sources to the weekly periodicals, and in early 2011, the first hurriedly published books appeared. Media and print publications detailed technical, environmental, and human stories about what took place, why it took place, and what responders did in addressing the situation. Many publications included diagrams depicting *Deepwater Horizon*, the wellhead, the drilling equipment, and the drilling process to educate readers on the specifics of deepwater drilling. To this, add the official reports, the most substantive of which was the 385-page report to the President by the National Commission on the BP *Deepwater Horizon* Oil Spill and Offshore Drilling (hereafter, the Oil Spill Commission): *Deep Water: The Gulf Oil Disaster and the Future of Offshore Drilling*.[57]

The majority of what has been written from all sources attempted to describe the technical aspect of the explosion, spill, and recovery; tried to explain the impact of the spill on Gulf residents, businesses, and the oil industry; or editorialized the spill's actors into one of three categories: hero, victim, or villain. Unfortunately, little of what has been written discussed intelligence support to the *Deepwater Horizon* response.

CAPT Erich M. Telfer

Lots of Daily Reporting, Little about Intelligence

Journalists largely ignored the intelligence-support aspects of the spill response. National and international daily newspapers published, and placed online, volumes of stories about *Deepwater Horizon*, illustrating the magnitude of the spill and reflecting the interest expressed by the public. Newspapers prominently reported the *Deepwater Horizon* story, from the explosion until BP stopped the oil flow in mid-July. The *Times-Picayune* of New Orleans had 585 articles on this topic between April 20 and July 15, 2010, when BP capped the well. In that same time period, the *Washington Post* posted 166 articles and the *Los Angeles Times* printed 135 pieces. Compare this with a search for articles on Afghanistan and the U.S. military efforts there, and the numbers are 42, 624, and 351 for each paper, respectively. The *Guardian* of London carried 144 articles about *Deepwater Horizon*. Closer to home, the Mexican daily paper *Reforma* included 36 articles between April 20 and July 15, 2010, with the term "*Deepwater Horizon*" in the piece. Despite the 1,066 pieces written about *Deepwater Horizon* from just these sources, very few of them discussed intelligence support to the response effort.

Online Sources

Online sources also reported widely on the spill. The *Coast Guard Digest*, an unofficial website that aggregates links to reports and comments on stories and issues involving the U.S. Coast Guard, carried 381 items pulled from media sources from when the *Deepwater Horizon* explosion first occurred until July 15, 2010. Out of the 381 pieces, just four of them touched on intelligence support issues and only six mentioned the lack of proper plans for such an explosion and spill.

Periodicals—Touching on Intelligence

By examining the *Deepwater Horizon* explosion and spill in greater depth, some weekly magazines brought the reader closer to matters touching on intelligence support to the response effort. The *New Yorker, Fortune, Newsweek*, and Economist ran several articles about *Deepwater Horizon* between April 20 and July 15, 2010.

New Yorker *Article*

The definitive periodical piece on the *Deepwater Horizon* explosion and spill is Raffi Khatchadourian's article in the *New Yorker* titled, "The Gulf War," which not only dissects the incident, but describes in depth the complete lack of preparation at the federal, state, local, and business levels for a spill of this magnitude. If one could read only a single article about the *Deepwater Horizon* incident to gain the broadest understanding, Khatchadourian's article would be it.

In his 24-page article, Khatchadourian describes the spill, the response, and, of special interest here, some aspects of intelligence support to the spill response. In particular, he describes the difficulties the responders had in finding where the oil drifted away from the wellhead, which impacted their ability to collect information on the oil and then skim it. Khatchadourian also discusses the shoreline cleanup and assessment teams who were disbursed daily to impacted areas to report on the oil they observed there.

Members of the SCATs described themselves to Khatchadourian as "intelligence officers for the cleanup."[58] In the parlance of the Intelligence Community (IC), the SCATs were conducting human intelligence (HUMINT), although this is a term not used in environmental response. Khatchadourian also talked about the lack of a plan to respond to the spill, an absence that extended to intelligence support. He wrote that the responders recognized the *Deepwater Horizon* explosion and spill as *the* worst-case scenario (for which there was no plan), and they began the tactical and operational response planning from that point forward.

Khatchadourian also addressed intelligence more directly. He quoted Ed Lavine, a chief federal scientist and oilspill responder going back to the 1970s, on how he viewed his key knowledge requirements: "Our mantra was: What got spilled? Where is it going? What's in its path?" In the language of the Intelligence Community, these are "priority intelligence requirements" (PIR) that the decisionmakers must have answered to prosecute the response. Again, "PIR" was not the term used, but the functionality—the form—is the same.

Fortune's *Favor*

Fortune's investigative piece, "BP: 'An Accident Waiting to Happen,'" describes how the wrong safety focus led to poor planning and incident response by BP.[59] The authors, Peter Elkind, David Whitford, and Doris Burke, told how a culture of individual versus process safety resulted in the *Deepwater Horizon* explosion and spill. Published in late January, 2011, the article described BP's history of safety challenges, the explosion and spill, the reaction of the public and the government, and the changes BP initiated in response to the incident. The article contains an illustration that compares the depth of the Macondo well to that of other wells, Mount Everest, and several skyscrapers. This illustration underscores the incredible distance covered by the *Deepwater Horizon* operation under water and earth, and why that complicated the response effort. In intelligence terms, this is also a priority information requirement.

Newsweek*'s Contribution*

Newsweek's most thoughtful article about the explosion and spill, "Black Water Rising," by Evan Thomas and Daniel Stone, described the explosion, spill, and government response, but was also open to misunderstanding. First, the authors claim that, when DHS Secretary Janet Napolitano declared this to be "a spill of national significance," this was "authorizing federal assistance to the region."[60] As will be explained shortly, a spill of national significance does not authorize federal assistance as is the case with a Stafford Act response (e.g., the response to Hurricane Katrina), but instead places a federal agency in charge of the response, as directed in the National Contingency Plan. In this case, the U.S. Coast Guard was given the role of the federal agency put in charge of making certain the *responsible party*, BP, covered the expenses and adhered to its legal responsibility to clean up the spill. Second, the article suggested that the oil was everywhere and coming ashore almost everywhere. This was not true, as the oil was actually rather difficult to find, away from the wellhead.[61] This highlights another priority information requirement mentioned in many of these articles: finding the oil. Remote sensing played a prominent role in locating, tracking, and classifying the oil, so responders could skim it and lay boom. By suggesting that the oil was everywhere, Thomas and Stone ignored the efforts and challenges of remote sensing in finding the oil. Finally, the piece reported that BP had an insufficient, poorly written response plan, despite the plan's approval by the federal government. This highlights another function of intelligence—support to planning.

The Economist*'s Take*

The *Economist's* coverage of the *Deepwater Horizon* explosion and spill focused on lessons to be learned from the spill, and the spill's impact in environmental, political, and economic circles. The May 6, 2010, edition included an article titled, "The Politics of Disaster," which speaks specifically about how government policy and regulation must gain better insight and understanding of what happened *for the future*. The magazine argues that disasters can be instructive, but that it is also easy to learn the wrong lessons.[62]

Despite the significant newspaper and weekly magazine coverage of the *Deepwater Horizon* explosion and spill, almost no space was devoted to intelligence support, with the exception of Khatchadourian's piece in the *New Yorker*.

Official Reports—Thin Soup on Intelligence

The U.S. government agencies involved in the response began publishing their after-action reports and "lessons learned" pieces in late fall 2010, including the official Presidential and U.S. Coast Guard studies.[63] The majority of these official reports focus on the lack of planning for a spill of the magnitude of *Deepwater Horizon*, and the absence of proper preparation in responding to such a spill.

The National Incident Commander's Report

Admiral Allen authored this 27-page report focused on "the critical strategic issues associated with the response."[64] The Admiral calls out a few items germane to the discussion here about intelligence support to disaster response. He describes his primary responsibility as promoting unity of effort across the entire governmental response. He also praises the Pollution Contingency Plan as giving him the discretion and freedom of action to respond to the spill. Additionally, however, Admiral Allen observes that the response constructs of the NCP and the NRF are different, and should be reconciled to improve federal effectiveness after an incident. He also comments on the challenges in establishing control of the airspace above and adjacent to the spill, and recommends that the example of the Air Operations Center established to support the DWH response be memorialized in doctrine for future spill response. This report was submitted to the Secretary of Homeland Security, Janet Napolitano, the first day of October 2010, and is one of the earliest official reports.

The Oil Spill Commission's Report

The Oil Spill Commission's report, titled *Deep Water*, is a lengthy, well-organized piece, published in January 2011, that covers many aspects of the explosion and spill, but is thin on addressing intelligence support to the response. As the foreword states, the report aims "to determine the cause of the disaster, and to improve the country's ability to respond to spills," as well as to recommend changes to the drilling industry to improve safety.[66] The commission members reached several conclusions, and pointed out that the ability of the federal government to respond to spills lags behind the risks associated with deepwater drilling, and that the government "must close" this gap by working with industry. Although not stated in the report, included in this "lag" is intelligence support to the responders. In addition, the report says that "neither BP nor the federal government was prepared to deal with a spill of the magnitude and complexity of the *Deepwater Horizon* disaster."[67] The report goes on to recommend that the "EPA and the Coast Guard should establish distinct plans and procedures for responding to a 'spill of national significance,'" and those should be based on industry's worst-case estimates of a spill.[68]

The Coast Guard's Unofficial (Official) Review

The Coast Guard chartered an interesting report about the *Deepwater Horizon* explosion, entitled *BP Deepwater Horizon Oil Spill: Incident Specific Preparedness Review* (ISPR), that touches on intelligence support. This *Incident Specific Preparedness Review* is not the official Coast Guard after-action report. The team that drafted the ISPR represented federal and state agencies, in addition to industry members. However, the Coast Guard commandant chartered the ISPR, endorsed it, and included an official cover memorandum directing Coast Guard service members to read it. One could be forgiven, perhaps, for taking the ISPR as a "Coast Guard view" of the incident, despite Admiral Papp's protestations in his cover memorandum. The authors of the report speak frankly, at one point criticizing the DHS for "severely restricting" the ability of the Coast Guard to release "timely, accurate information" to the public.[69] This drew the attention of Fox News's Mike Levine, who, in his March 28, 2011, article, "While Slowing BP Oil Spill, Administration Slowed Flow of Information Too, Claims Coast Guard Report," appeared confused about the focus of the ISPR.[70] In the article, Levine quotes an unnamed Coast Guard spokesman as telling Fox News that the ISPR "does not reflect the views of the Coast Guard."[71] Fair enough, but what do the authors of the review say about intelligence support to the *Deepwater Horizon* response effort? They do not say much—but they do say more than most.

The ISPR Outlines Problems with the Common Operating Picture

The *Incident Specific Preparedness Review* speaks to intelligence support mainly when describing the challenges of developing the common operating picture (COP) and the lack of knowledge management. A COP provides accurate, timely, and relevant information to assist operational and strategic decision-making.[72] In terms of the intelligence cycle, the COP would fit into production and dissemination, because a COP gets the information (and intelligence) to the decisionmakers in a useable form. Initially, no unified COP existed, and information was fragmented over many systems with "approximately 10 different GIS databases being used to track spill response information."[73] The bandwidth at the Unified Area Command (UAC) complicated matters, with responders struggling even to send e-mails.[74]

In the end, the responders at the Incident Command Posts (ICP) developed their own COPs, as did BP and other nongovernmental responders. No national-level COP existed for more than a month, according to the ISPR. (My research suggests it was closer to two months, as will be described later.) The lack of a single, accessible COP frustrated Admiral Allen's call for unity of effort, including intelligence support. That support consisted, in part, of NGA providing imagery enhancing the "tactical decision making of critical resources movements on a real-time basis."[75] Proprietary information, information-handling caveats (such as "For Official Use Only"), and agency firewalls delayed the use of a single COP.

Ultimately, the federal, state, local, and private responders settled on a University of New Hampshire product used by NOAA, called the Environmental Response Management Application (ERMA). A GIS-based system, ERMA permitted the UAC to organize and display information from multiple systems onto one product with layered strata. The genesis and use of ERMA during the *Deepwater Horizon* response will be described in Chapter 5.

The Coast Guard ISPR outlines the necessity of a standard, exercised COP *before* an incident. Such a COP would permit responders from multiple government and private organizations to manage and share information quickly within a commonly understood construct.

The review also highlights the deficiencies in knowledge management systems, which would include the ability to produce and share intelligence, and information management systems shortfalls. Recommended is a standard report template that "captures the oil spill response essential elements of information and other key metrics" to meet responders' information needs.[76] Essential elements of information (EEIs) would be part of a properly managed and led planning process and intelligence cycle. They define what the decisionmakers (tactical, operational, and strategic) need to know. The EEIs then shape the intelligence-collection plan by defining requirements, and so on. The authors of the review made an insightful point here, but they did not go far enough in describing the lack of intelligence plans and functions.

Lest the reader suppose that the ISPR engages intelligence support to the *Deepwater Horizon* response thoroughly, I draw attention to the third page in Admiral Papp's cover memo. Following Coast Guard protocol, the last portion of the memo contains the distribution list: the major headquarters offices and operational commands that are to receive the memo for action. The Coast Guard office of intelligence (CG-2) is not listed. Staff officers put considerable thought into memorandums for the commandant's signature, and the omission of CG-2, while perhaps unintentional, may signify a gap in understanding that intelligence *even has a role* in disaster response— and, even worse, that CG-2 might not be cognizant of the ISPR and its findings.

On Scene Coordinator Report

In September 2011, the On Scene *Coordinator Report* was published by the U.S. Coast Guard to document the response to the spill. The federal on scene coordinator is the lead federal authority for conducting the response to a spill of national significance under the direction of the national incident commander. This report is the Coast Guard's most comprehensive *Deepwater Horizon* after-action report promulgated since the spill. It is well organized and thorough, and includes an expanded executive summary that briefly describes the content of each chapter. While the report does not address intelligence directly, it does comment on several topics discussed in this work. Chapter 2, "Command and Control," highlights the misunderstanding among state and local governments over the authorities that the federal government exercised during a response to a spill of national significance. Chapter 3, "Operations," describes the challenge in locating and tracking the oil, especially that which had moved away from the well sight, as well as deployment and use of the boom. Chapter 3 also discusses skimming operations, Department of Defense support to the response, and the shoreline cleanup assessment techniques. Chapter 6, "Logistics," covers the vessels of opportunity, including command and control issues, aviation safety, aviation coordination, aviation tasking, and the Air Force Aviation Operations Center at Tyndall, Florida. This chapter praises the joint and interagency efforts run out of the Air Force base at Tyndall, especially in integrated remote sensing, strategic planning, and mission support liaisons. Chapter 10, "Communications," discusses in great detail the common operating picture concept and ERMA, which the responders adopted to manage and lead response operations. Despite the excellent organization of this report and the wealth of details it contains, the report is often written in the passive voice. For example, one passage reads, "During the early states of the spill, determining who collected what data in what formats, and how to access it, was challenging."[77] The passive voice permits a point to be made without directing fault. I suspect this writing style was pragmatically used to avoid offense.

Books—Recent and Detailed (Except about Intelligence)

The comparative recentness of the *Deepwater Horizon* explosion and spill means authors have had little time to study, put into perspective, and write about this event of national significance. Much has been studied and written about the disaster, but the field of thoughtful work narrows considerably, if not completely, when searching for literature specifically about intelligence support to disaster response. This author was neither surprised nor dismayed by the paucity of books focused specifically on intelligence support to unanticipated events of national significance. However, a brief examination of some texts that describe disasters and chaos may assist in appreciating the nature of a response to an event of national significance. The publishers began selling the first books about the *Deepwater Horizon* explosion, spill, and response in early 2011.

Disaster Books—The Value of Being Prepared

There are a considerable number of books that examine disasters and disaster preparation. This author discovered no books focused specifically on intelligence support to disaster response. Of all the books on disasters, a recurring theme is how the unexpectedness of a disaster (whether manmade or natural) complicates and confounds the response efforts. As there was no plan in place for how intelligence would support a spill of national significance, a few of these books are worth mention.

The Black Swan

Nassim Nicholas Taleb's 2007 philosophy book *The Black Swan: The Impact of the Highly Improbable* discusses man's inability to comprehend the inevitability of disasters and therefore to plan for them.[78] For the purpose of this study, the interesting material in *The Black Swan* concerns an inability of individuals and organizations to plan and prepare for catastrophes. Instead, Taleb explains, individuals and organizations expect and plan for what most often occurs. Events do occur outside of the "normal" spectrum; worst-case scenarios do take place, but humans do not sufficiently expect them.

The Unthinkable

Intelligence professionals at the tactical level supporting disaster responders must also understand that people, including the responders, do strange things in a catastrophe. People can act bravely, stupidly, with compassion, or with malaise when terrible events happen around them or to them. This is the theme of Amanda Ripley's book *The Unthinkable: Who Survives When Disaster Strikes—and Why*.[79] Ripley studied several cases of disaster—from the 9/11 attacks to Hurricane Katrina—in an attempt to understand people's reaction to catastrophe. She argues that professionals are not immune to the effects of an unanticipated disaster and may still exhibit strange reactions after a disaster. Finally, Ripley suggests that government agencies have a duty to consider and make plans to address worst-case scenarios.

Perils of Progress

In his book *Perils of Progress: Environmental Disasters in the 20th Century*, Dr. Andrew Jenks describes how large-scale, manmade disasters are normal and recurring in modern times.[80] This book will be examined in greater detail later in this work.

The Deepwater Horizon Books

October 2010 saw the first of what will likely be a gaggle of books about *Deepwater Horizon*. However, publication of these books may have been too close to when the *Deepwater Horizon* explosion and spill happened to give the material a balanced treatment. Some words of caution are warranted, especially when reflecting on Barbara Tuchman's warning that writing about an event too soon after its occurrence may sacrifice detachment. But Tuchman preceded this by writing that "facts are history whether interpreted or not" and the authors of the newly published books illustrate many facts of interest.[81]

Disaster on the Horizon

Bob Cavnar's book *Disaster on the Horizon: High Stakes, High Risks, and the Story Behind the Deepwater Well Blowout* (published October 2010) critically examines the *Deepwater Horizon* explosion and spill, and argues that the incompetence of BP and the federal government set the conditions for the catastrophe and then prevented a proper response. In his description of the *Deepwater Horizon* explosion and the initial weeks following it, Cavnar explained the panic and lack of certainty about what had happened in the explosion and the extent of the oilspill. His discussion of the dispersants Corexit 9500 and Corexit 9527, which BP "sprayed almost indiscriminately"[82] and introduced under water at the leak site, raises the question of what impact, if any, the dispersant had on the ability of remote sensing to detect oil for the response effort. Cavnar also faults the U.S. government for its lack of preparation and planning for worst-case scenarios of oilspills when responders use dispersants, booming, and skimming. This lack of planning may have extended into intelligence support to such a spill, although Cavnar does not say as much.

The oil industry likely will not care for Cavnar's work, however, which ends on a sour note. (See Steve Mufson's February 13, 2011, *Washington Post* review of Cavnar's book, as well as the two other books discussed in this section.[83]) Cavnar's book is anti–oil industry and anti-BP. He speaks of BP as a global business tyrant focused on making money, eschewing transparency, and ruining lives. Cavnar goes so far as to suggest that the U.S. government "has become dependent on oil royalties to fund its mammoth expansion and war fighting" overseas.[84] Cavnar also suggests that Admiral Allen did not comprehend the complexity of the disaster and that he was merely a mouthpiece for BP.[85] While he raises some interesting points and speaks with experience and authority, his bias taints the reader as to Cavnar's motivation and brings into question his judgment on the complicated matter of the *Deepwater Horizon* spill and response. Unfortunately, Cavnar neither addresses nor directly discusses remote sensing, imagery, and geographic information systems in his description of the spill response.

CAPT Erich M. Telfer

In Too Deep

Stanley Reed and Alison Fitzgerald's book *In Too Deep: BP and the Drilling Race That Took It Down* describes BP's "meteoric rise" and then its sudden fall. The authors focus on BP and its senior leaders, drawing parallels between the avarice, ambition, success, and decline of the company with those qualities in the senior leaders. In closing, the authors discuss suggested changes to the oil industry and deep-sea drilling to avert another event such as *Deepwater Horizon*. Some of the suggestions include better preparation for responding to oilspills, which would certainly (although the authors do not say so) include intelligence.

Blowout in the Gulf

In their book *Blowout in the Gulf: The BP Oil Spill Disaster and the Future of Energy in America*, William R. Freudenburg and Robert Gramling look at how the *Deepwater Horizon* explosion and spill occurred. By drilling and operating in "increasingly dangerous waters" without adequate vigilance and forethought, BP set the conditions for the Macondo well blowout and spill.[87] The authors also describe the inadequate "fantasy documents" that pass for response and contingency planning for accident and spill response.[88] This is important because the authors condemn not only BP and industry, but the federal government for approving these inadequate plans. As my work argues, in part, the lack of intelligence planning in these response plans bears examining.

Summary

The U.S. government response guides, "lessons learned" documents, academic work, media reporting on the spill, and recently published books are insufficient to address the performance of intelligence during the *Deepwater Horizon* response. At best, they point to a lack of understanding about the definition, role, and function of intelligence in responding to a spill of national significance. The federal response plans give too much autonomy to decisionmakers in the organization of intelligence within a response, and pay too little attention to the true goals of intelligence: warning decisionmakers and helping them make better decisions. Two graduate papers highlighted the shortcomings of ISR support to disaster response and stated that, even though intelligence contributed to the response to Hurricane Katrina, tactical responders were not well served. *Deep Water* was detailed, but very little of the discussions touch on intelligence support to the spill response. Newspapers extensively covered the spill, but rarely, if ever, mentioned intelligence. A few periodicals did discuss the role and function of intelligence in response to *Deepwater Horizon*, but most recently published books do not. Available after-action reports say little about intelligence, but do discuss the general unpreparedness of the federal government in reacting to a spill of national significance.

In summation, very little literature exists that focuses on intelligence support to disaster response, and even less to supporting a spill of national significance. To answer whether and to what degree intelligence supported the *Deepwater Horizon* spill, I therefore sought out those who participated in the intelligence effort, and the decisionmakers served by intelligence.

Chapter 2—
Methodology: Authentic Answers to Authentic Questions

"The historian must make do with what he can find."

—Barbara Tuchman

The purpose of this project is to understand the process of intelligence support to the *Deepwater Horizon* response. What happened? Instead of attempting to describe national-level intelligence systems (satellite imagery) in terms of capabilities and collection, I sought to understand the relationships between the decisionmakers at the strategic and operational levels and the intelligence supporters. I concerned myself, as well, with the tactical responders and their perception of intelligence support to their mission of finding and skimming oil. What did intelligence support do? And, more importantly, what did intelligence support not do?

The methodology of this project involved systematically collecting data on the intelligence gaps and vulnerabilities in the support to the *Deepwater Horizon response*. I mainly accomplished this task with a thorough review of publicly available documents (highlighted in Chapter 1) and interviews with participants in the crisis response. After this initial data collection, I took the results and made inferences regarding a potential plan for intelligence support to future spills of national significance.

Data Collection

Because of the paucity of available writing on the *Deepwater Horizon* response, I determined that interviews would be the most valid means of data collection for learning about intelligence support to the *Deepwater Horizon* response. By early 2011, the existing literature included several government and official reports, in addition to a growing volume of commercially published material. These works proved highly instructive about the history of deepwater drilling, the events leading up to the explosion, the explosion itself, and the following containment and cleanup. As described in Chapter 1, however, even as the U.S. government and publishers printed and released this material, little of it engaged the issue of intelligence performance and support to the *Deepwater Horizon* responders.

So, in order to collect the information needed for the study, I decided to seek out those individuals most involved in intelligence support to the response effort. I wanted to hear authentic voices from firsthand experience. Ultimately, I conducted 34 interviews. I focused my interviews and questions around the time from the explosion on April 20, 2010, until mid-July, when the well was capped. Intelligence officers continued to support Coast Guard and cleanup operations after BP stopped the oil leak, but, as will be described later, these intelligence functions changed little, if at all.

Interviews by Organizational Role

I organized the interviews into groups based on the role the interviewee played during the response effort. By organizing the interview subjects into groups, I could customize the interview questions to best address the groups' experience and understanding of the intelligence support to the response. The groups were liaisons, tactical responders, Incident Command Post staff, senior decisionmakers, academics, and imagery analysts. The interviewees were U.S. Coast Guard members (active and reservists), Air National Guard members, civilians, and academics.

Liaisons

Liaisons work as their service's representative within or as part of another organization's functions. Out of the liaisons with whom I spoke, few had specific direction as to their role vis-à-vis their parent and supporting organization. Mostly, the liaisons served as communications conduits between their parent and supporting organization, in addition to working as part of the staff. Liaisons were often brokers or handlers of intelligence, but usually not formally trained intelligence officers.

Tactical Responders

Tactical responders consist of U.S. Coast Guard cutter commanding officers who were skimming oil or controlling other surface and air assets or both. These commanding officers constitute a core customer group for intelligence, in that their mission included locating skimmable oil. Oil proved very difficult to locate from the surface (i.e., from the deck of a ship). Because of this, the Unified Area Command, the operational-level command, and the Incident Command Posts (the tactical-level commands) used "remote sensing" or satellites to help direct the skimming operations.

Incident Staff

The *Incident Command Post staff* comprises personnel assigned from many agencies. Out of the hundreds of people who worked in the ICPs, those I interviewed were Coast Guard members who worked from the level of deputy incident commander to within the cutter task force under the Field Support Element. These Coast Guard staffers were uniquely positioned to observe and experience how intelligence supported the spill response.

Decisionmakers

Three senior *decisionmakers*, Coast Guard admirals all, accommodated me for interviews regarding their experience with intelligence support to the spill response. They were Admiral Thad Allen, Rear Admiral Paul Zukunft, and Rear Admiral Peter Neffenger. Admiral Allen served as the national incident commander from April 29, 2010, while he was still the U.S. Coast Guard commandant. Admiral Allen retired in June 2010, but remained the national incident commander until Rear Admiral Zukunft took over the spill response on October 1, 2010. Admiral Zukunft assumed the role of the federal on scene coordinator (FOSC) on July 10, 2010, where he led the collaboration of the many federal, state, and local responders to the spill. Finally, Rear Admiral Neffenger served as the deputy national incident commander directly under Admiral Allen. I sought the perspective of these senior-level decisionmakers because the Department of Homeland Security designated the Coast Guard as the lead agency in the Deepwater Horizon response, and these officers led that response. These men were also key customers of the intelligence support effort.

Academic Researchers

I was fortunate to visit with and receive the assistance of seven academics during my research. I met professors of history and geology, as well as professional educators in the fields of geospatial information systems and disaster management. I sought a broader perspective from these interviewees based on their research and time in their fields. Louisiana State University and the Earth Scan Laboratory were particularly helpful in guiding my understanding of GIS and the limitations of satellite imagery in responding to a spill of national significance.

CAPT Erich M. Telfer

Imagery Analysts

Imagery (including remote sensing, geospatial information systems, and satellite and airborne collection) made up the largest portion of the intelligence response effort. I conducted interviews with several personnel who participated in the imagery support. I found these interviews valuable in understanding what initiatives enhanced the spill response, as well as what frustrated this response. The vast majority of the information garnered from imagery analysts during the interviews was for attribution, but some was not. Regarding information passed to me "not for attribution," the reader should understand that I made certain that the interviewee had placement and access to comment professionally on the matter at hand. While a researcher must weigh the value of information "not for attribution" against the protocols for citation, I determined that, in most instances, I could include the information and honor the interviewee's request for anonymity.

Conduct of the Interviews

I contacted 42 people and asked to speak with them about their experiences regarding *Deepwater Horizon*. Out of this group, I conducted 34 interviews. I held interviews both in person and via phone. I met with several interviewees in Washington, DC, and also traveled to Portsmouth, Virginia, Tyndall Air Force Base in Florida, and Louisiana State University in Baton Rouge, Louisiana, to research and meet with interview subjects. Eight people whom I contacted were unable to discuss their experiences regarding intelligence support to *Deepwater Horizon*. Out of those eight, one was on assignment away from her work office and could not be reached; four set up multiple appointments with me, but failed to follow through with the meeting; and three subjects would not speak with me. The shortest interview lasted about 35 minutes and the longest went for several hours, over an entire afternoon.

I captured all interview data via literal transcriptions from digital recordings or typed interview notes. I digitally recorded four of the interviews, including those of each flag officer, and wrote direct transcripts from those conversations. For the balance of the interviews, I typed my notes. In all cases, I sent a copy of the transcript or typed notes to the interviewee and asked him or her to examine my work for clarity and completeness. Some interviewees did not respond to the request for review of my notes, and I have accepted their lack of a response as agreement with the material. After receiving the interviewees' responses, I included the information in my research material.

One Question Can Elicit a Thousand Words

I developed the interview questions to seek out information unavailable from the literature review. The questions were open-ended and intended to encourage the interviewee to speak with candor about his or her understanding and experience with intelligence support to the *Deepwater Horizon* response. I did not use a standardized questionnaire for academic interviews, as I tailored those discussions specifically toward the subject and his or her area of expertise. Likewise, I customized each leader's questionnaire based on the position of the interviewee. Despite using tailored questions, I weaved some similar themes throughout for consistent data collection, and to compare and contrast responses among the groups.

A brief explanation behind the methodology of the common questions will underscore the data-collection plan.

Question 1: Please briefly summarize your career background, experience, and education.

I asked this question to understand the perspective of the interviewee regarding the response effort, and as a way to get the interviewee talking. Firsthand accounts of any incident possess inherent problems, including tunnel vision (overemphasis on one issue to the exclusion of other issues), misremembering (i.e., forgetting), and cultural perspective (responders with different backgrounds and professional experiences view events differently). By asking the interviewees to describe themselves, I was able to adjust "on the fly" for these problems, and also was able to weigh the interview responses. For example, an interviewee with years of disaster response experience likely will have a perspective that is different from that of responders whose first incident was *Deepwater Horizon*. Even in informal, conversational settings, interviewees may be nervous, especially when discussing the greatest oilspill in U.S. history. I found that by asking the interviewees about themselves, they were put at ease and spoke more comfortably.

Question 2: When did you arrive and depart from the response operations?

The timeframe of most interest to me was from the *Deepwater Horizon* explosion until BP capped the well: April 20–July 15, 2010. Some interviewees supported the response after July 15, and I included their input in my research.

Question 3: What role did you play in the *Deepwater Horizon* response?

In some interviews, I framed this question differently by asking interviewees for their title in the response position. The idea here was to find out if the interviewee actually did the job or filled the role that the interviewee thought he or she had been detailed to do (as opposed to having a title and job description, but being tasked to do other missions). Also, I felt it was important to know how the interviewee viewed his or her contribution to the response.

Question 4: What were the information needs of the responders/decision-makers you supported?

This question sought to elicit a key point about intelligence support—what did your customers need to know? For the tactical responders, I asked them to identify their information needs and to explain how they communicated those needs and to whom they communicated them.

Question 5: How were those needs met?

For intelligence support personnel, I asked how they or their organization performed in supporting decisionmakers with intelligence. For decisionmakers, I asked how intelligence supported them as a decisionmaker. This question also addressed the functionality of the intelligence and the performance of the intelligence cycle.

Question 6: If those needs were not met, how did you communicate that up the chain?

Here, I wanted to know how responsive intelligence was in addressing the gaps when the decisionmakers did not receive the intelligence they had requested. What was the interaction between intelligence officers and decisionmakers? How well did that interaction work?

Question 7: What information/intelligence needs remained unanswered?

Intelligence is not a magic "all-knowing" answer box. No matter how well intelligence performs, decisionmakers will rarely, if ever, have all the information they want and need. However, I felt it was important to map the outer edges of what intelligence was ultimately unable to answer during the *Deepwater Horizon* response. The natural follow-on question therefore would be, "Should intelligence have been able to provide that information? And if yes, how so?" As an aside, I found in my interviews with intelligence professionals that few senior-level decisionmakers knew how to ask for intelligence.

Question 8: About what are you most proud of in your support of *Deepwater Horizon*?

I also asked about what went well during the response. Again, giving interviewees the opportunity to reflect on their hard work and contribution to the response proved quite fruitful in data collection. When I first formulated these questions, I felt this question was not going to be very successful. Instead, I learned valuable information by asking interviewees to describe things that went well. I often asked them to tell me the good news they shared with their family and friends upon returning from the *Deepwater Horizon* response.

Question 9: What area or issue gave you the most friction and frustration in your support of *Deepwater Horizon*?

This question is the corollary to the pride question. My initial research before conducting the interviews indicated that many intelligence efforts did not yield success, or yielded limited success, and I wanted interviewees to tell me what had vexed them the most. This forced interviewees to categorize and prioritize their frustrations and gave me the data to compare responses against a single measure—what was the most significant failure of intelligence in the spill response?

Question 10: What recommendations would you make, based on your experience, regarding the information/intelligence support you received?

I wrote this question to solicit ideas on improving intelligence support to future spills of national significance. This question is cheating, a little bit, because I asked the interviewees to tell me how they would fix the frustrations they had just described. I found the answers to be enlightening, however, because those who participated in the intelligence support effort were keenly aware of the limitations of intelligence during *Deepwater Horizon*, and all agreed that intelligence could have performed better. The interviewees provided concise advice on how to improve intelligence support to future spills of national significance.

Question 11: On a scale of one to ten, with ten being the best, how would you rate the performance of intelligence in support of the *Deepwater Horizon response?*

This question forced the interviewees to rate intelligence support from their perspective. This question is an oversimplification of a complicated matter, and several of the interviewees expressed frustration in trying to answer it. I did not think this was a particularly strong question, but the findings proved interesting. For example, strategic and operational decisionmakers had a much different opinion of the performance of intelligence, compared with the opinions of the tactical responders. In addition, several interviewees did not give a single number as a response, but qualified their answers to span periods of time during the disaster or between different intelligence functions. These numbers hardly constitute a scientific set, but they do offer an interesting insight into their overall perceptions of intelligence performance.

Question 12: Are there any other matters regarding your role in the *Deepwater Horizon* response that you would like to discuss that we may not have covered?

This was a catchall question to afford the respondents a chance to speak generally about *Deepwater Horizon*. Several interviewees were interested in making strong points to me regarding their experience with *Deepwater Horizon*. Had the earlier questions not permitted them to address these points, I found they did so for this last question.

Chapter 3— Insufficient Intelligence Plans Hampered Response

"Pause for serious thought is not a habit of governments."[90]

—*Barbara Tuchman*

It is safe to say, based on the many interviews I conducted, that the lack of an existing intelligence plan to support a spill of national significance hampered the *Deepwater Horizon* response effort. This chapter reviews the main findings from the literature review and interviews, and includes a discussion of the lack of preparation for an event of the magnitude of *Deepwater Horizon*. There was no plan because the government and private organizations involved in the *Deepwater Horizon* response had failed to envision the actual "worst-case" spill scenario. Thus, the plans that did exist were insufficient to deal with a deepwater blowout and did not contain sufficient discussion and guidance for intelligence support. This chapter also underscores the absence of command and control of the intelligence effort that hampered the work of the intelligence officers who deployed to support the response effort. Finally, I recommend actions to address the shortcomings identified in this chapter.

Lack of Imagination Hampered Planning and Preparation

None of the organizations studied for this research were prepared for a worst-case scenario spill in deep-sea drilling, and this lack of preparation extended to intelligence support plans. Government and BP contingency plans barely mentioned intelligence support to a spill, including geospatial information systems or satellite tracking of the spill. In addition, the U.S. government practice of reviewing those plans did not include key responders, such as the Coast Guard, in reviewing the plans. Because BP did not envision a true worst case, the company failed to develop methods to track the spill, including the flow rate of oil from the blown-out well.

BP's "Worst Case" Was Not Bad Enough

Organizations do not plan well for large-scale, manmade disasters even though they are reoccurring events, according to Andrew Jenks, the author of *Perils of Progress.*[91] In the case of oilspill planning, both the U.S. government and the oil companies it regulates develop plans to respond to spills. The primary planning methodology for that response is the National Incident Management System. For an oilspill, the responsible party pays for the response as directed by the Oil Pollution Act, but the U.S. government directs that response. In actuality, the response is more of a partnership or "alliance," as Joel Achenbach, author of *A Hole at the Bottom of the Sea: The Race to Kill the BP Oil Gusher,* described it regarding *Deepwater Horizon.*[92] When a spill occurs, the U.S. government and the responsible party come together and implement the plans they have written.

Dr. Jenks argues that some institutions lack "disaster imagination," a concept Jenks uses to describe the lack of ability to plan for a worst-case scenario, or even to envision what may occur.[93] While the past may be instructive, Jenks comments that governments and organizations display "historic amnesia" about what took place in previous disasters, and fail to pull out the pertinent facts and conclusions to influence planning.[94] When governments and organization fail to plan for the worst-case scenarios, then they also fail to plan the intelligence support to assist in the response to the catastrophe.

"Get Ready . . . !"

Nobody was prepared for a deep-sea well blowout. Media, literature, and government after-action reports have all cited a lack of preparation and planning for a spill of national significance resulting from deep-sea drilling. The Oil Spill Commission said that "neither BP nor the federal government was prepared to deal with a spill of the magnitude and complexity of the *Deepwater Horizon* disaster."[95] According to Bob Cavnar, the federal government and industry were poorly prepared and failed to procure and properly deploy dispersants, boom, and oil-skimming assets.[96] The Oil Spill Commission's report also adds that "oil spill response planning across the government needs to be *overhauled.*"[97] (Emphasis added.) As mentioned previously, William Freudenburg and Robert Gramling, in their book *Blowout in the Gulf,* called the BP and government response plans written before the explosion "fantasy documents."[98] The U.S. government guides on disaster response do not address intelligence, at least not in a meaningful manner, and neither did BP's response plan.

BP's Spill Response Plan

The *Regional Oil Spill Response Plan—Gulf of Mexico*, written by BP, addresses intelligence sparingly. This 582-page plan was written as a set of "easy-to-follow instructions . . . in the event of a release of a product" in the Gulf of Mexico.[99] The plan does list geospatial information systems within the duties and responsibilities of the situation unit leader, but the focus is on understanding or tracking the current status of units (ships and aircraft) responding to the spill. The BP plan is not a traditional intelligence cycle. Later in the plan, GIS is listed within the organization of the incident management team.

Page 7 of the plan, however, stresses the importance of tracking the movement of an oilspill and predicting its trajectory. While this may seem to be a normal engineering task, it could also be considered a classic intelligence mission: finding something, determining what it is, and figuring out what it is doing. Satellite imagery is included as a tool for tracking the spill. This is important, because a large spill, contrary to the media reporting with *Deepwater Horizon*, is difficult to track. Unfortunately, in addition to its intelligence gaps, the Oil Spill Commission's report says (on page 84) that the BP plan lacked detail and was not serious.[100]

U.S. Failure to Review BP Plan

In addition to an insufficient BP contingency plan, U.S. government inaction prevented better review of that plan. The federal government directs that oil industries include, among other things, a plan for determining the trajectory of the spilled oil. Even when these plans were current, which BP's were not before the *Deepwater Horizon* explosion, the Mineral Management Service (MMS) rarely shared them with other federal agencies with expertise in oilspill response—including the Coast Guard.[101] The lack of exposure to the plans meant that the Coast Guard, as the key response agency, was caught flat-footed and unaware of what contingencies BP expected and what they planned to employ, assuming the contingencies accurately anticipated the magnitude of a spill.

Lack of Command and Control

U.S. government disaster guidelines for spills of national significance do not detail a single agency as the intelligence lead. Because of this, no one agency effectively stepped forward to lead the intelligence effort in supporting the *Deepwater Horizon* responders. Despite several federal agencies deploying intelligence officers to support the spill, unity of command among the intelligence responders remained elusive. Frustrating the work of the intelligence officers was the absence of an intelligence plan to support a spill of national significance. Problems with command and control persisted even after Admiral Allen tasked the 601st Air and Space Operations Center in mid-June 2010 to manage the airspace and the remote-sensing operation.

The First Rule of "Having a Plan" is Having a Plan

The U.S. government disaster response guides do not establish unity of command within the intelligence function. As discussed in Chapter 1, the intelligence function may reside in the planning or operational sections, or as a standalone section within the Incident Command System, at an Incident Command Post, or at a Unified Area Command. The Federal Emergency and Management Agency (FEMA) conducts remote sensing (imagery) and collection management (writing and validating priority intelligence requirements) during Stafford Act responses. In such a case, FEMA would direct and manage the intelligence support effort wherever it may be placed within the Incident Command System.

No Intelligence Plan for SONS

However, a spill of national significance does not require a Stafford Act response, because it is governed under the National Contingency Plan, and FEMA did not take a role in the intelligence support to the *Deepwater Horizon* response. Instead, several agencies, including the Department of Homeland Security, the National Geospatial-Intelligence Agency, the Coast Guard, and the Department of Defense, sent intelligence officers to support the effort and to lead the intelligence operations. Unfortunately, because the national disaster guidance is silent on the matter of how intelligence support should aid in a spill of national significance, the intelligence responders had difficulty determining who was supposed to drive the intelligence effort. One responder commented that, "going into the fight," there was no government-wide plan for integrating imagery into a planning process and operational support for a spill of national significance.[102] Neither DHS nor the Coast Guard had a plan for using intelligence to support the spill responders.[103] When a remote-sensing concept of operations (CONOPS) was needed, intelligence staffers used the Stafford Act Remote Sensing CONOPS as a model.

Another participant in the *Deepwater Horizon* response, Stan Gold of the Coast Guard, explained how the National Response Framework lacks guidance on where to place remote sensing within the organization.[104] He went on to say that "we don't have a national response plan, we have a national response framework. So it's all coalition."[105] Coalitions are tricky things. It can be difficult to determine who is in charge, especially when responding to a crisis. Coalitions are also difficult to manage when the intelligence officers come from numerous and disparate agencies (i.e., U.S. Coast Guard, NGA, NOAA, and DoD) that do not regularly train for intelligence support post-crisis. Add to this the incredibly complex, multilayered, geographically diverse nature of the *Deepwater Horizon* spill, and building and maintaining an intelligence support "coalition" became an exceedingly difficult task.

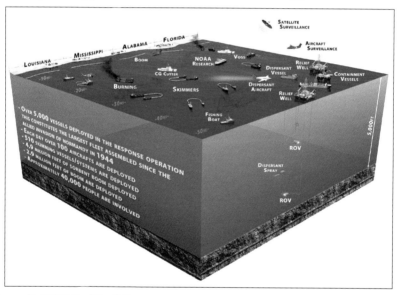

Figure 5: USCG, "Building ICS Organization"

Source: Captain Meredith Austin, USCG, "Building ICS Organization During the Deepwater Horizon Response," National Fire Protection Association, undated, accessed September 1, 2011, *www.nfpa.org/assets/files/metro%20chiefs/MetroAustin2011. ppt*, slide 3.

CAPT Erich M. Telfer

Geospatial Information Systems (GIS) and Human Intelligence Support to *Deepwater Horizon*

Federal agencies deployed intelligence officers to the Unified Area Command and liaisons to the Incident Command Posts (ICPs) to support the spill response. For example, DHS Secretary Janet Napolitano directed the entire DHS Interagency Remote Sensing Coordination Cell (IRSCC) to lead that agency's intelligence support effort to the *Deepwater Horizon* response. On May 1, 2010, the IRSCC first deployed a handful of staff members to Roberts, Louisiana, and then to New Orleans when the Unified Area Command (UAC) relocated there. Once in New Orleans, the IRSCC members took direction from Coast Guard staff at the UAC on behalf of the National Incident Commander and in coordination with the responsible party, BP. The National Oceanographic and Atmospheric Administration also had staff in the UAC who contributed to the intelligence effort by suggesting the priority intelligence requirements.[106] The Department of Defense contributed as well when staff from the National Geospatial-Intelligence Agency and Air Force North sent liaisons to the UAC and the ICPs.

The majority of intelligence support personnel worked in imagery, remote sensing, or geospatial information systems. The Coast Guard sent intelligence officers to assist in coordinating the intelligence processes, intelligence production, and tactical intelligence support to the skimming task forces. The shoreline cleanup and assessment teams (SCATs) also served an intelligence function, although employment of the SCATs is a specific part of oilspill response doctrine, and its members are not traditional intelligence officers.[107] Not including the SCATs, no fewer than 100 intelligence professionals took part in supporting the full range of the spill response.[108]

Department of Defense GIS Operations

The Department of Homeland Security (DHS), the National Oceanographic and Atmospheric Administration (NOAA), the National Geospatial-Intelligence Agency (NGA), the Department of Defense (DoD), and BP all sent members to conduct imagery and remote-sensing work in support of the response. In early May 2010, the U.S. Air National Guard 601st Air and Space Operations Center entered the fight also, to assist in imagery and remote-sensing management. NGA representatives arrived in early May as well. During a Stafford Act response, the DHS Interagency Remote Sensing Coordination Cell leads imagery intelligence as the collection manager. In this capacity, imagery planning, collection, analysis, and production are coordinated through IRSCC. This creates unity of effort and unity of command during a chaotic incident response. However, U.S. government guidance is silent on what agency is designated as the collection manager in response to a spill of national significance, which is one reason why so many imagery and remote-sensing intelligence officers deployed.

Lots of Staff, But No One in Charge

Lieutenant Colonel Remso Martinez, USA, described a leadership vacuum within the UAC in early May regarding intelligence.[109] This occurred even though there was sufficient staff at the UAC, and the main intelligence function, the IRSCC, was nominally in charge of the intelligence effort by direction of the DHS secretary. Stan Gold described the sincere desire of the intelligence professionals to do their work well and contribute to the response effort.[110] However, interviews with 15 ICP and UAC staff who worked in and around the intelligence section clearly indicate that a leadership vacuum did exist. Martinez said, "I saw a lot of good people doing good work. But I saw a lot of people doing busy work that didn't yield success."[111] Intelligence professionals with good intentions and work could not overcome the lack of a scheme on how to successfully support the response effort.

> *Intelligence professionals with good intentions and work could not overcome the lack of a scheme on how to successfully support the response effort.*

Attempts Were Made to Develop Command and Control

According to information gathered in interviews, several unsuccessful attempts were made to establish unity of command among the intelligence responders. Just as nature abhors a vacuum, military members abhor the absence of unity of command. The first two weeks after the explosion on April 20, 2010, were chaotic. No organization existed among the intelligence agencies responding and no agency was designated to lead. The Coast Guard Eighth District out of New Orleans, led by Rear Admiral Mary Landry, attempted to provide the intelligence support to the spill response, but the Eighth District intelligence staff was quickly overwhelmed by the scope of the problem and the burdens of trying to maintain their regular, non-spill intelligence work.[112]

The intelligence effort shifted to the Unified Area Command. According to Stan Gold, "All hell broke loose on about the 26th or 27th [of April] . . . when we realized how bad things were."[113] Gold also explained that nobody within the Intelligence Community preparing to support the response in late April and early May 2010 knew who was in charge, who was to be supported, and who was supporting the effort. Among the intelligence agencies and responders, Gold added that "everybody blew it in the beginning."

Reactive Intelligence Slowed the Response

The lack of an intelligence support plan meant that intelligence staffs were reacting to data calls, as opposed to driving the intelligence cycle. Even worse, this lack of sufficient unity of command contributed to several near-catastrophes as aircraft saturated the space above and around the spill site.

With No Command and Control, There Is Only Reaction

The intelligence staffs were reacting. For example, within the planning section of the UAC, the intelligence officers prepared weather briefs and answered short-fuse data calls in the first month and a half after the explosion, instead of driving an iterative intelligence cycle to support strategic, operational, and tactical decisionmakers. Admiral Allen said that his staff, including the intelligence element in the planning section, had difficulties in answering the data calls coming in from the White House and the Department of Homeland Security. "The number one driver [of information] was an insatiable demand for instant data by the [DHS] Secretary and the Deputy Secretary that was just driving everybody crazy," said Admiral Allen.[114] This was especially the case regarding inconsistent facts reported by the media and federal agencies to Washington about the flow rate of the spill and the amount of boom deployed. As previously discussed, determining the flow rate proved extremely difficult. Determining how much boom had been deployed also became a daily data point in briefings and information passed back to Washington.

Failure to Develop Priority Intelligence Requirements

The intelligence personnel tried to develop intelligence requirements, but the lack of command and control hindered their efforts.[115] In early May, the

IRSCC members drafted priority intelligence requirements (PIRs), but they were unable, initially, to have the UAC validate the requirements (i.e., confirm that those requirements were the information needs of the UAC).[116] The PIRs were not developed in an iterative process, where the decisionmakers explained their information and intelligence needs to the staff. Instead, a member of the IRSCC was told by a Coast Guard captain, "Whatever NOAA wants [for imagery], that is our requirements."[117] The IRSCC wrote the priority intelligence requirements with input from NOAA staff members, but without UAC guidance. The UAC did not approve the PIRs in May, although their use became the practice for intelligence collection. The PIRs were approved in late June, but they were not disseminated to the ICPs.[118]

BP Confounds Efforts to Collect Intelligence

From an intelligence-planning and command-and-control perspective, the constant presence of BP representatives as the responsible party confounded unity of effort. In the case of the Air Force 601st, the unit is used to control incident awareness and assessment (IAA) assets in both domestic and overseas combat operations, and tasking those assets as required by the intelligence collection plan. In a SONS response, however, the responsible party pays for all operations, which essentially amounts to approval authority. If BP refused to pay for a flight, then the parent agency was unlikely to conduct the flight. According to Captain Colin Washburn, USAF, 601st AOC, this added an additional layer to all decisionmaking regarding IAA tasking, and made the 601st liaison officers and staff feel they lacked sufficient control to address the priority intelligence requirements.[119] BP had to be consulted on every issue involving intelligence collection by government assets.

Conclusion: The Requirement for a Plan and Command and Control

As is certainly obvious in hindsight, an intelligence plan to support a spill of national significance likely would have helped address several of the issues raised in this chapter. "An intelligence plan is necessary," observed Rear Admiral Neffenger, the deputy incident commander. "It can't eliminate the fog, but [intelligence] gives you a better grasp of the situation."[120] A plan to coordinate intelligence, establish intelligence unity of effort, and manage the intelligence cycle across the response effort was necessary. When I asked Lieutenant Colonel Martinez why an intelligence plan was lacking, he responded, "People seemed comfortable reacting. Planning requires hard, careful thought."[121] Even though the intelligence agencies and staff that supported *Deepwater Horizon* lacked a plan, they responded nonetheless to provide information and intelligence to the strategic, operational, and tactical decisionmakers.

Findings and Recommendations

1. **DHS should direct the Coast Guard to develop an intelligence support plan for a SONS scenario.**

The DHS should develop an intelligence support plan for the next spill of national significance. DHS should require that the response to a spill of national significance contain an intelligence section that stands apart and is independent from the other sections (operations, planning, logistics, and administration). Since the Coast Guard would likely be the lead agency in responding to such a spill, the Coast Guard commandant should be responsible for developing the intelligence plan. The plan should envision a worst-case spill, regardless of the cause. A worst-case planning scenario would help prepare a more sufficient intelligence response and, at the same time, acclimate senior decisionmakers to the scope of the potential catastrophe. In developing the plan, the Coast Guard should seek input and support from the academic community and coordinate the plan with the applicable federal agencies that likely will provide intelligence support in the spill response, including NOAA and the Department of Defense. This intelligence support plan should be incorporated into overarching spill response plans and tested, as the *On Scene Coordinator Report* recommends, in order to prove useful during an actual spill.[122]

2. **Parameters of an Intelligence Support Plan**

The intelligence support plan for a SONS should detail the command relationship within the Incident Command System. The intelligence section would report to and receive guidance from the incident commander. (See Chapter 1, Figure 4.) The intelligence section of the plan should establish direct lines of command and control and of relationships among the intelligence agencies supporting the response. The plan should be flexible and written to accept the participation of state, local, tribal, and private groups in responding to a spill.[123]

The intelligence component should anticipate a significant GIS/remote-sensing role in the response, most likely provided by a Department of Defense entity, and include mechanisms on how GIS will support the incident commander and the response. The stakeholders should periodically review the plan for accuracy and relevance to make certain it remains appropriate and useable, and addresses changes in technology. Finally, the plan must be practiced during SONS exercises. It must include the participation of the intelligence officers and staff who would respond to an actual spill.

Chapter 4—
The *Deepwater Horizon* Intelligence Cycle: Spinning But Off Balance

> "No one is so sure of his premises as the man who knows too little."[124]
>
> —*Barbara Tuchman*

The intelligence cycle used in the *Deepwater Horizon* response evolved over two months from simple reaction to managing day-to-day intelligence challenges. In other words, no intelligence support design or standard operating procedures existed before, during, or after the crisis. First of all, responders at the strategic, operational, and tactical levels knew what they wanted to know, but they communicated those requirements only informally. Despite the informality and disorganization, a significant amount of data was collected, primarily through remote sensing. However, analyzing those data streams turned out to be more difficult because of hardware and software deficiencies. Finally, producing and disseminating the intelligence proved ineffective at the tactical level, although decisionmakers at the operational and strategic levels gave higher marks to intelligence support to their efforts.

Missing from the Beginning: A Basic Intelligence Cycle

To briefly review from Chapter 1, the intelligence cycle is a process that describes the basic functioning of intelligence. The cycle consists of determining the intelligence requirements, organizing a plan to collect those requirements, collecting the information, analyzing the information to add value and context, and then producing and disseminating the intelligence.

Step 1: Establishing Intelligence Requirements

From the strategic to operational and down to the tactical levels, decisionmakers involved in responding to the *Deepwater Horizon* disaster knew what they wanted to know. They wanted to know where the oil was, where the oil was going, where the boom was that had been laid out to intercept the oil, and where the ships were in responding to the spill. Admiral Allen described these requirements as the "vital signs" that the DHS secretary and White

CAPT Erich M. Telfer

House officials wanted to know daily.[125] The federal on scene coordinator, Admiral Zukunft, echoed Admiral Allen's information requirements, as did the Department of Homeland Security, Coast Guard, and National Geospatial-Intelligence Agency staff working in the Incident Command Posts. Intelligence requirements were developed by the staffs, but not via a formal process, and the requirements were not validated in an orderly manner. (In fact, some requirements were never validated.) According to Commander Robert Jensen, USCG, the tactical responders (i.e., the commanding officers of the Coast Guard cutters who were skimming oil and directing the hundreds of "vessels of opportunity") wanted to know "where is the oil now and where will it be tomorrow?"[126]

Response Structure Hindered Intelligence Requirements

The nature of the response structure to a spill of national significance hindered the development of intelligence requirements. A SONS intelligence and remote-sensing response, unlike a Stafford Act response, has neither a designated *location* nor *function* within the Incident Command System, as discussed in Chapter 3.[127] Because of this, the initial intelligence staffers who arrived from the Coast Guard and the Department of Defense to assist at the Incident Command Posts (operational level) and the Unified Area Command (strategic level) to support the response could not easily be integrated into the overall staff model. No location existed into which the intelligence staff could coalesce. Moreover, the function of the staff was in question. In early May, BP worked alongside the incident command staff to drive the information and intelligence requirements. According to Captain Caesar Kellum, USAF, 601st Air and Space Operations Center, while this appeared to be "fairly successful," there was little overarching collaboration among the intelligence staff and agencies participating.[128] One participant described the confusion that existed within the UAC staff, especially regarding the BP employees, about how intelligence and remote sensing could assist in the response effort.[129]

In addition to a lack of clear location and function for the intelligence staff, there was no script for developing intelligence requirements—those things the decisionmakers needed to know. The first DHS intelligence officer working for the Interagency Remote Sensing Coordination Cell arrived at the Unified Area Command in Roberts, Louisiana, on May 1. Throughout the spill response, the IRSCC maintained an intelligence officer in Roberts, and then in New Orleans, to help coordinate intelligence prioritization and the intelligence cycle. Along with Coast Guard and NOAA staff, the IRSCC officer wrote the first set of commander's critical information requirements (CCIRs) for the federal on scene coordinator (FOSC) during the first week of May. But those CCIRs were neither universally understood nor even acknowledged at the UAC. As late as early June, several senior staff at the UAC did not even know about the intelligence requirements.[130] At a level below the FOSC, the four Incident Command Posts[131] coordinated, developed, and managed their own intelligence, primarily remote sensing, by building requirements that were neither validated nor tied to the requirements of the FOSC.[132] Even though they were never formally validated, these requirements still drove planning and collection throughout the response operation.[133]

Early PIRs Developed by the IRSCC

The priority intelligence requirements listed in Figure 6 were developed by members of the DHS IRSCC during the first week of May 2010, about 10 days after the *Deepwater Horizon* explosion. The intent with the initial PIRs was to link collection platforms (aircraft, satellites, etc.) to the intelligence needs to collect the information decisionmakers wanted.

PIR 1: Location depth, thickness, density, total volume, fprward edge, and projected movement of the British Petroleum Oil spill. Report current and projected locations, encroachment on shoreline, and estuaries, impact to waterways, fisheries, and wildlife.

PIR 2: Locate and quantify the active oil containment booms off the coast deployed in reponse to British Petroleum oil spill.

PIR 3: Locate and quantify and displaced oil containment booms off the coast deployed in response to British Petroleum oil spill.

PIR 4: Provide baseline overhead imagery depicting the US coast, its estuaries, and coastal waterways pior to the encroachment of oil from the British Petroleum oil spill for the purposes of change detection. Imagery should be collected since May 8, 2009 (no more than one year oil).

Figure 6: Unified Area Command Priority Information Requirements (PIRs) from May 18, 2010.[134]

Source: U.S. Department of Homeland Security, daily brief, "Federal Remote Sensing Situation Report – British Petroleum Oil Spill Response, May 18, 2010," slide 5.

By early July, the UAC had expanded the PIRs from Figure 6 and added detailed essential elements of information to the list. Identifying and articulating the PIRs was a valid first step in determining the information needs of the decisionmakers.

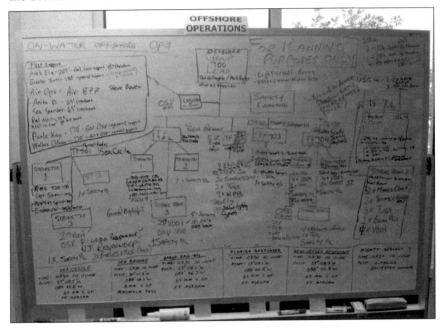

Figure 7: Operations Center photo

Source: USAF 601st Air and Space Operations Center photo taken by and courtesy of Maj Collin Washburn, ISR Operations Team Chief.

An officer in the Air Force 601st Air and Space Operations Center provided the photo in Figure 7, the "On-Water Offshore Ops" taken on May 17, 2010. The white board depicts the various task forces (listed as "TF") and subordinate units that were searching for and skimming oil. On the upper-right section of the board was written "For Planning Purposes Only." The May PIRs previously listed were designed to focus efforts in finding and describing the oil, as well as determining where the oil was going. Once obtained, that information was to be used to plan skimming operations and to direct the government, contract, and privately owned ships. But the oil was not easy to find.

Locating and Describing the Oil

Locating the oil proved challenging. Although this may seem impossible with thousands of barrels of oil gushing out of the wellhead daily, finding and tracking the oil four or five miles away from the wellhead was "a huge issue," according to Stan Gold.[135] Gold also reported several instances when a Coast Guard cutter was within 10 yards of an oil slick, but still could not see it.[136] Oil coverage was not uniform and did not cover a large area "as was perceived," according to the *On Scene Coordinator Report*.[137] A significant observation effort was needed to locate the oil. This required aircraft at altitudes of at least 500 feet all the way up to satellites. Once the oil was found, the information had to be communicated to the incident command posts and the Coast Guard cutter commanders, so that they could direct surface skimmers, apply dispersants, track the oil's movement toward ecologically sensitive areas, and keep the public informed of the spill.

Coast Guard cutter crew members are not uniformly trained and practiced in looking for oilspills. As Commander Robert Jordan, USCG, put it, "Oil does not act like a person in the water," meaning the Coast Guard knows how to search for people (and boats and aircraft) in the water, but the oil proved more difficult.[138] Oil patches broke away from the main slick, streamers developed where the oil would float along like long red-orange ribbons, and oil also traveled under the surface and came up a dozen miles from the wellhead.[139]

Graphics built by NOAA and NGA that were passed to strategic decision-makers in Washington and the media did not depict the broken-up, unpredictable movement of oil. Instead, as shown in Figure 8, the graphic depicted a large, solidly shaded area where any oil *may* have been located. Even with a key at the bottom of the graphic, this gave the impression that the entire shaded area of the Gulf of Mexico was covered with oil—a fact later lamented by Rear Admiral Peter Neffenger.[140] In his comments to me, Neffenger explained that the depiction left a false impression with the public and White House staff members that oil was easily located and covered an area exponentially larger than it did. PIR 1 indicated what information the decisionmakers wanted regarding the oil, but the intelligence effort could not deliver on the tasking.

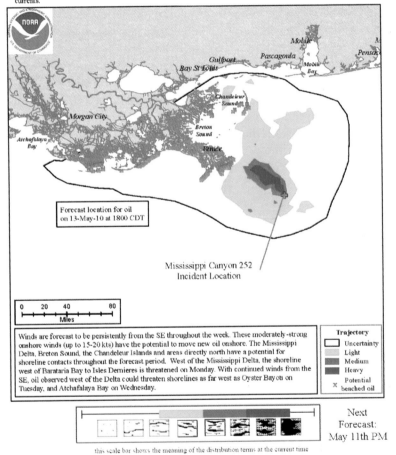

Trajectory Forecast
Mississippi Canyon 252

NOAA/NOS/OR&R
Estimate for: 1800 CDT, Thursday, 5/13/10
Date Prepared: 1700 CDT, Monday, 5/10/10

This forecast is based on the NWS spot forecast from Monday, May 10 PM. Currents were obtained from the NOAA Gulf of Mexico, West Florida Shelf/USF, Texas A&M/TGLO, and NAVO/NRL models and HFR measurements. The model was initialized from Monday afternoon overflight observations, satellite/aerial imagery and analysis provided by NOAA/NESDIS and Transport Canada. The leading edge may contain tarballs that are not readily observable from the imagery (hence not included in the model initialization). Oil near bay inlets could be brought into that bay by local tidal currents.

Figure 8: Trajectory Forecast Mississippi Canyon
Source: National Oceanic and Atmospheric Administration, "Trajectory Forecast Mississippi Canyon 252," accessed June 4, 2011, *http://deepwaterhorizon.noaa.gov/ bookshelf/1963_TMF72-2010-05-10-1700.pdf.*

Locating and Determining the Status of the Boom

Locating and tracking the boom deployed to intercept the oil became a critical intelligence requirement, because it became a measure of the effectiveness of government response. "Boom was the coin of the realm," Commander Richard Timme, USCG, observed.[141] BP, the federal government, and state governments all deployed thousands of miles of boom during the response. At its height, there were 13 million feet (2,462 miles) of boom deployed.[142] Boom was a means of showing the public that the government was responding. In particular, the deployment and location of boom was extremely important to parish presidents in Louisiana.[143] By early July 2010, tracking the boom became just as important, if not more so, as tracking the oil.[144] Unfortunately, in the confusion and chaos of the first six to eight weeks after the explosion and sinking of the *Deepwater Horizon*, the federal and state governments, in addition to BP, had no method of tracking the boom once deployed. Rear Admiral Neffenger, the deputy incident commander, explained that "much of that boom, at any given moment, was off station."[145] "Off station" means the boom was not where the ICPs wanted it because wind, tides, and currents had moved the boom after it had been placed. The *On Scene Coordinator Report* echoes Neffenger's comment that "most of the booming was counterproductive" but viewed as "necessary as oil approached the shore."[146] The *On Scene Coordinator Report* does not, however, answer my question, "necessary to whom?" I suspect that the boom placement was necessary to manage public opinion and perception of the spill response.

Oil-containment booms are floating lines used to contain, deflect, or corral oilspills. Boom is not normally used on the open ocean. It is best used in a confined area, such as a harbor, as opposed to being deployed for miles in open water. Almost any weather (e.g., wind and waves) will disrupt the integrity of the boom and move it, permitting oil to pass through the barrier.[147] Not surprisingly, the wave and wind action in the Gulf often moved the boom from its initial deployment locations. This is why a critical information requirement became finding and tracking the boom.

Figure 9: Containment boom during *Deepwater Horizon* response, Pensacola, Florida.

Source: Photo by U.S. Coast Guard.

Locating and Tracking Vessels of Opportunity

While not included as a PIR in the IRSCC list, locating and tracking the volunteer and contracted surface response ships was another critical information requirement. The incident command posts organized all of the surface ships into task forces, with a Coast Guard cutter in the lead. The cutter directed subordinate cutters, skimmers, and vessels of opportunity (VOOs). The Coast Guard cutters and other federal vessels, such as NOAA ships, were easily tracked because they have systems onboard that monitor and transmit their locations. However, in order to direct skimming operations, the Incident Command Posts and the commanding officers of the Coast Guard cutters needed to know where all the other ships were as well.

Unfortunately, tracking the thousands of vessels of opportunity that BP and the states employed to spot and skim oil proved extremely difficult.[148] These VOOs were a diverse and plentiful flotilla that included more than 10,000 vessels.[149] The ships ranged from thousand-ton offshore supply vessels with state-of-the-art dynamic positioning equipment, to shrimp boats, and all the way down to canoes.[150] The expanded after-action report, "Deepwater Horizon Response Surface Operations: A Case Study Prepared by Participating WLB Commanding Officers," edited by Commander John Kennedy, USCG, explained that communicating with the VOOs generally occurred via VHF-FM radios. However, the bandwidths were crowded and most of the surface task forces ended up developing their own communication plans on the fly, using channels not normally employed.[151] The basic problem was that few of the VOOs had transponders that would enable the Coast Guard cutters to track their location and movement. The eight Coast Guard 225-foot buoy tenders (WLBs) and multiple patrol boats tried to "loosely coordinate" the VOOs, but the span of control and communications "slowed information flow and the fusion of data for tactical decision-making."[152]

Tracking the VOOs was also important for conducting Coast Guard skimming operations. About 1.5 million gallons of oil were recovered by Coast Guard cutters from early May to late July 2010. Ninety-five percent of that oil was recovered by the eight seagoing buoy tenders. The buoy tenders[153] could fill their oil storage tanks in a matter of hours if the oil encountered was thick. To continue skimming, however, the WLBs needed to tow tank barges alongside in which to store the skimmed oil, or they needed to offload the oil to a contracted VOO barge. If the cutters could not offload the oil, the Coast Guard crews would be unable to work.[154] The ability to track, locate, and contact the contracted VOO barges was also a critical information requirement for the WLBs and the ICPs.

The ICPs also had significant challenges in managing the VOOs. As Haley Barbour, then Mississippi's governor, testified at the House Committee on Oversight and Government Reform's June 2, 2011, hearing, he learned that the ICPs and the Unified Command could neither communicate with the VOOs, nor track their location when a "significant amount" of oil reached his state's barrier islands in early June without warning.[155] Since there was no command and control system to effectively share information with and direct the VOOs, there was no way to efficiently use the VOOs to report on the location of oil and assist in skimming. To address this, the state of Mississippi purchased radios in June and established a system that facilitated communication among the VOOs, the ICPs, the Coast Guard surface forces, and patrolling aircraft.

Baseline Imagery: Before the Spill

NOAA provided pre-spill imagery to the UAC in response to PIR 4, as did commercial imagery sources and some academic labs, including Louisiana State University's Coastal Studies Institute. The UAC used the pre-spill imagery as a baseline to observe the impact of oil by studying the differences in imagery of the shores, coasts, and estuaries. NOAA's pre-spill imagery was from early June. However, that imagery had gaps and "skips" (areas not covered by the pre-spill imagery catalog) that NOAA addressed through government and commercial imagery suppliers even as oil approached the shoreline in early May, according to David Gisclair, technical assistance program director of the Louisiana Oil Spill Coordinator's Office.[156]

Evaluation of PIRs

The absence of a system to gather and process intelligence requirements hampered the spill response, even though the decisionmakers from Admiral Allen down to the commanding officers of the Coast Guard cutters knew what information they needed to do their work. Even when intelligence officers from the Coast Guard, NOAA, and the IRSCC drafted priority intelligence requirements, those requirements were not disseminated across the response effort.

The first PIR (finding the oil) was more difficult than anticipated. This was exacerbated by misleading graphics passed to the media and senior decisionmakers in Washington depicting oil covering the entire northern Gulf of Mexico. If oil was present everywhere, why search for it? Next to the oil, locating boom and tracking the vessels of opportunity were also top priorities. Locating the boom showed effort in the spill response and was relatively successful. On the other hand, efforts to locate and track the VOOs to support skimming operations were less successful.

But determining and agreeing on priority intelligence requirements was just the first step. The next step was to develop a remote-sensing concept of operations to answer the PIRs. This was done, but not fully implemented.

The Remote-Sensing CONOPS

Just a week after the oilspill, the IRSCC staff in Washington, DC, drafted a remote-sensing concept of operations for the *Deepwater Horizon* response. The object of the draft CONOPS was to formalize remote-sensing planning and operations. The IRSCC staff used the existing remote-sensing concept of operations for Stafford Act catastrophes as a model in developing the *Deepwater Horizon* remote-sensing CONOPS.

Write It and They Will Sign It (or Not)

Receiving formal validation of the early draft of the remote-sensing CONOPS, however, proved more difficult, according to a memo written by the UAC's remote-sensing coordinator (RSC). As detailed by Lieutenant Chris Lucero, USCG, in his August 3, 2010, memo, "USCG Remote Sensing Coordinator Statement: Non-BP Signature and Loss of *Deepwater Horizon* Remote Sensing Concept of Operations Document," the intelligence staff successfully conveyed to Rear Admiral James Watson, the federal on scene coordinator (FOSC), the importance of approving the CONOPS. Watson signed it on July 9, 2010. From there, the staff forwarded the signed CONOPS to Douglas Suttles, the FOSC's BP counterpart, for his review and signature. The UAC staff also forwarded the CONOPS to the IRSCC on July 13 for use on an interim basis because the RSC expected BP's approval without delay. However, despite assurances from Coast Guard and BP legal staff that the CONOPS was sound and no legal issues were impeding the approval, Suttles refused to sign the document. Then, on July 27, David Randall, a BP executive, explained to the UAC staff that, since the *Deepwater Horizon* well had been capped on the 15th and the amount of oil found in the water was significantly reduced, BP did not need to sign the CONOPS. When the UAC staff attempted to retrieve the CONOPS, Randall explained in an e-mail on August 3 that BP had lost the original CONOPS signed by Rear Admiral Watson.[157] The originally signed document was never recovered. This was an unfortunate event because the remote-sensing CONOPS was the first formalized attempt at organizing remote-sensing intelligence support to a spill of national significance. Had the document been signed and implemented, it would have set a precedent for naming the Coast Guard as the lead agency and the business (BP) as the responsible party. Moreover, the

CAPT Erich M. Telfer

CONOPS would have set a model for future remote-sensing collaboration in response to a spill of national significance.

BP Refusal to Sign

According to Lieutenant Lucero in a follow-up phone interview, Douglas Suttles's refusal to approve and sign the CONOPS may have had less to do with plugging of the well than with what the CONOPS would have meant to BP autonomy.[158] BP maintained its own remote-sensing coordinator, who managed a parallel effort to those of the IRSCC, the UAC staff, and the Air Force 601st Air and Space Operations Center. The BP remote-sensing coordinator directed surveillance flights that were collecting imagery to support BP needs. BP did not coordinate these flights with the IRSCC or the 601st. In addition, BP did not consistently share the imagery collected from its contracted remote-sensing flights. Had Suttles signed the CONOPS, BP would have been bound to coordinate its remote-sensing flights along with the larger effort overseen by the IRSCC and the 601st. This would have decreased BP's autonomy and increased its requirement to share information.

Failure to Establish a Formal CONOPS

As explained in the *Deepwater Horizon* Unified Area Command's document, "Remote Sensing CONCEPT OF OPERATIONS: Deepwater Horizon Response," the goal of the remote-sensing CONOPS was "to provide information to contain the spread of oil and recover the spilled oil," which would also mitigate the environmental impact.[159] The purpose of the plan was to define the capabilities of remote-sensing operations to facilitate the response, to build the Incident Awareness and Assessment (IAA, or ISR in DoD terms) structure within the UAC, and to define the UAC remote-sensing management team (RST) processes. The RST would be an interagency body of remote-sensing professionals that would enable the UAC "to plan, coordinate, acquire, analyze, publish, and disseminate situational knowledge."[160] In addition, each ICP involved in remote sensing would have an RST assigned to its planning section. According to the CONOPS, the UAC would establish overall remote-sensing requirements, but would delegate the development and planning of remote sensing to the 601st Air and Space Operations Center. The UAC could then task remote sensors that fell under UAC command, and request that tasking be assigned to remote sensors not under UAC command.

66

Step 2: Intelligence (Dis)organization: Collection Planning

Intelligence organization was haphazard in the initial weeks after the explosion of the *Deepwater Horizon*. When the intelligence cycle is running smoothly, decisionmakers validate intelligence requirements, and then staff members develop a plan to collect, analyze, produce, and disseminate the intelligence. *Deepwater Horizon* was not an optimal situation, however, and intelligence organization suffered. While Admiral Allen had a general idea of his information needs, "it was four to six weeks to get everything down and nailed" with an organized process for planning collection.[161] Admiral Allen was so concerned about overburdening and duplicating information and intelligence planning and collection that he instructed his staff officers to question and challenge him if they perceived he was imposing additional intelligence requirements on them.[162]

Never as Bad as the First Time

Intelligence organization for a spill of national significance was a unique event. *Deepwater Horizon* was the first designated spill of national significance since the *Exxon Valdez* spill in 1989, when the state of domestic intelligence capabilities and support was much less. "We were plowing new ground here," Stan Gold explained. In particular, no one knew how to integrate all of the information coming in from national-level remote-sensing assets; BP remote sensing; or state, local, and foreign-contracted imagery collection of the spill.[163] Prosecuting the intelligence cycle in the initial weeks after the explosion was therefore the result of largely informal information sharing among participants, not the result of executing predetermined intelligence plans.[164]

One DHS intelligence officer explained that the most challenging aspect of supporting the response was coordinating and planning imagery collection.[165] According to Commander Russell Dash, USCG, even as aircraft and satellites were detailed to collect imagery of the spill, there was neither a plan nor a system for matching flights with collection sensors on the aircraft and satellites.[166]

Moreover, several interviewees said there was a bureaucratic fight at the UAC over who was in charge of imagery collection and platforms. Several agencies jockeyed for position in early May to be designated as the lead in remote sensing, including the National Geospatial-Intelligence Agency, the Coast Guard, NOAA, and the Defense Department. As a result, none of their efforts were in concert.[167] In the absence of an integrated intelligence plan, as previously described, agencies acted parochially and sought out the information *their* staffs needed to respond to the agencies' decisionmakers. For example, Lieutenant Colonel Martinez described how a senior BP official flew around of his own volition visually searching for oil in a contract aircraft. The BP official argued that this was the best way to find the oil, by simple air searches and reporting the oil via voice communications.[168] A more thoughtful, ordered approach to planning intelligence would likely have helped build a more efficient means of collecting the information that the BP executive wanted.

The intelligence organization finally materialized in early July 2010. Up until this point, the Coast Guard intelligence liaisons at the 601st Air and Space Operations Center, the intelligence staff at the Unified Area Command, and the Interagency Remote Sensing Coordination Cell staff worked long hours to ensure more efficient remote-sensing operations. The IRSCC deployed staff to the *Deepwater Horizon* response from May 1 until August 19 (110 days), working 16 or more hours a day, seven days a week. Many of the intelligence liaisons, BP staff, and ICP staff had worked hundreds of hours together and were better acquainted by early June. This familiarity, based on personalities and personal interaction, enhanced intelligence planning and coordination and helped address some of the shortcomings caused by lack of a formal intelligence planning structure.[169] (As documented in Lieutenant Commander Dietrich's "The Eyes of Katrina," this was also the case during Hurricane Katrina.)[170]

Step 3: Collection

The story behind intelligence collection for *Deepwater Horizon* is largely about airspace control. Since the majority of intelligence required and collected to support the *Deepwater Horizon* response was done via remote sensing, airspace control became the central issue. Unfortunately, organizing and managing the collection platforms was haphazard through May and early June 2010. When flight safety became an issue, Admiral Allen asked the Air Force to take over airspace control, and this greatly increased flight safety as well as intelligence collection.

UNLIMITED IMPOSSIBILITIES

Flight Safety Issues

Several near air-to-air collisions motivated a change in the airspace management over the spill area. Despite the fact that the Federal Aviation Administration (FAA) established temporary flight restrictions days after the spill and
enlisted a Customs and Border Protection (CBP) P-3 to help monitor air traffic, flight safety remained a paramount concern.[171] On June 16, Admiral Thad
Allen asked the Department of Defense for an air tasking order, so that the Air
Force could take control of the airspace over the spill. The order was effective
in bringing some order to the airspace, and it also improved intelligence collection, even though several agencies did not participate in the tasking order.

The aircraft that saturated the *Deepwater Horizon* response area after the spill
created a dangerous situation. Federal, state, local, BP, and media aircraft
combed the area around the spill site, along the shores, and the area in between.
Even after the Air Force took control of the flights, from mid-July to mid-Au-
gust, U.S. government aircraft alone flew 390 missions covering 1,048 hours.
Most of these aircraft flew according to visual flight rules (VFR), meaning the
pilots conducted their own flight avoidance. The helicopters often flew "see and
avoid," meaning the pilot managed his own detection and avoidance of other
aircraft. In addition, the aircraft searching for oil and supporting media coverage did not fly in predictable patterns and did not follow predictable routes.
Not all of the aircraft reported back to, or even coordinated with, the ICPs, as
recounted by Captain Kellum in his unclassified brief, "Incident Awareness and
Assessment (IAA) Support to Deepwater Horizon (DWH)."[172]

So many aircraft operating in close proximity to each other without centralized control created a dangerous situation. There were eight near mid-
air collisions involving aircraft responding to and supporting the *Deepwater
Horizon* effort in late May and early June.[173] Despite this, senior-level Coast
Guard decisionmaker Rear Admiral Peter Neffenger argued that still more
aircraft were needed, especially from the shoreline to about 20 miles out, in
order to spot oil.[174] But the absence of control over the airspace caused more
than just safety problems.

Negative Impact on Remote-Sensing Capability

The lack of flight coordination across the response effort hampered intelligence
collection. According to Stan Gold, "There was absolutely no coordination

69

at the [operational] level or at the [tactical] level of the various aircraft and satellites that were used to provide information in support of planning efforts" to support the response.[175] This lack of flight coordination meant that there was no systematic method for collecting information about the location, type, and extent of the oil. This resulted in decentralized control, execution, and (unorganized) intelligence collection. When asked about the lack of a plan to support airborne intelligence collection, Admiral Allen commented, "This gets back to us not having a structure" to incorporate intelligence into remote sensing.[176] According to Commander Dash, inefficiencies existed until mid-June 2010 in matching aircraft with sensor packages along with priority information requirements.[177] Because aircraft and sensors were not coordinated across the response effort, different flights overlapped the same area multiple times in a brief period, while there was a "complete lack of coverage" in other areas.[178] Because the PIRs were not well understood, agencies tasked and dispatched flights to collect data supporting their own information needs and requirements. Admiral Allen not only needed to control the aircraft above and around the spill to prevent a collision, but he also needed control to better collect information about the spill. By mid-June, Admiral Allen realized that he needed an air tasking order.

Implementing the Air Tasking Order

The U.S. military uses an air tasking order (ATO) as part of a greater plan to guide air operations. An ATO is tactical direction promulgated every 48 to 72 hours, and encompasses the specific aircraft, missions, flight paths, and timelines. Only aircraft listed on the ATO are authorized to fly. Supporting the ATO are the Airspace Control Order (ACO) and any Special Instructions (SPINS). The ACO defines the operational area, communications plans, flight routing, radio contact procedures, and identification friend/foe (IFF) codes. The SPINS include guiding principles of the operation, such as the commander's intent, the mission objectives, and the rules of engagement. Most importantly, a single entity (usually the Combined Air Operations Center) controls these plans and enjoys complete unity of command. In addition to flight safety, the ATO designates intelligence-collection missions in support of the priority information requirements or the collection plan, and designates aircraft and sensor packages to most efficiently and effectively collect the information.

From the *Deepwater Horizon* explosion until early June, the response effort lacked an ATO. There simply was no coordination of aircraft across the response. According to interviews, the near collisions took place because no air tasking order was in place. And another interviewee suggested the ATO would not have been developed save for the near collisions.[179]

"I Need to Take Control"

On Tuesday, June 15, President Obama returned to Washington, DC, from a visit to the Gulf Coast. Flying on Air Force One along with the media, security, and staff was Admiral Allen. As recounted by Joel Achenbach in *A Hole at the Bottom of the Sea*, the former Coast Guard commandant sat chatting with an Air Force steward when the President sat down and asked Allen, "How's it going?"[180] "I need to take control," Allen replied.[181] He asked Obama's permission to take over the airspace above the spill. The President directed Admiral Allen to do what he thought needed to be done. "There are no do-overs," the President said, and Admiral Allen understood exactly what he meant.[182] "I got the idea [to take control of the airspace] from Haiti," Admiral Allen said, referring to his experience in the Coast Guard response to the Haiti earthquake. He made the comments to me in a March 2011 interview.[183] Regarding *Deepwater Horizon*, Admiral Allen commented that after his discussion with President Obama, he called the Chairman of the Joint Chiefs of Staff, Admiral Mike Mullen, and the chief of staff of the Air Force, General Norton Schwartz, as well as the United States Northern Command (USNORTHCOM) commander, Admiral James "Sandy" Winnefeld, to ask about the Air Force taking control of the airspace. Admiral Allen was told that it would be done the following day.

A Pivotal Point: The Air Force Steps In

Admiral Allen described the establishment of the ATO as the "pivotal point" in responding to the spill.[184] Many of the federal, state, and local responding agencies flying remote-sensing aircraft had been "doing their own thing for 20 years" and had never worked in a coordinated manner.[185] The following day, as part of taking over the airspace, Admiral Allen informed the responders in a lengthy e-mail that they would now manage the response as a "three-dimensional battle space" and move away from a traditional spill response approach.[186] Admiral Allen's observation indicates that, in addition to overwhelming BP and the nation's ability to respond, *Deepwater Horizon* also stymied the Coast Guard's own methodology for disaster response. Fortunately, the 601st Air and Space Operations Center was able to inject process into the remote-sensing intelligence collection, and this finally yielded success.

Issues Resolved: Enter the 601st

The 601st Air and Space Operations Center conducts both overseas and domestic intelligence support. Staff of the 601st are trained in Intelligence, Surveillance, and Reconnaissance (ISR), which, according to U.S. Air Force's *Intelligence, Surveillance, and Reconnaissance Operations: Air Force Doctrine Document 2-9*, is designed "to provide accurate, relevant, and timely intelligence to decisionmakers."[187] "ISR" is the term used in Department of Defense missions, especially combat support operations, but when supporting U.S. domestic disaster response, the same mission is usually described as "incident awareness and assessment." The function is essentially the same, but several interviewees were insistent that the term "ISR" should not be used in conjunction with domestic operations.

The 601st Air and Space Operations Center became the Aviation Coordination Command within the Incident Command Structure, and, as such, received guidance from the UAC and in return supported both the UAC and the ICPs. But the *Deepwater Horizon* response was not a military operation, so the 601st did not enjoy full unity of command. Even when the ATO was in place in mid-July, the 601st was able to control only a few flights.[188] Some non-military agencies—NOAA in particular—did not participate in the formal ATO, choosing instead to inform the 601st about flights, as opposed to subordinating or even coordinating their flights and missions.

Even though it is a Department of Defense unit, the 601st was the perfect outfit to establish an ATO because, as Lieutenant Colonel Susan A. Romano wrote in her *Air Force Print News Today* article "Deepwater Horizon Airspace Activity Now Coordinated at 601st AOC," it was "no stranger to responding to natural disasters. In the wake of the earthquake that devastated Haiti in January 2010, the 601st AOC was tasked to assist with airspace deconfliction and air flow in and out of the Port-au-Prince airport, while maximizing the efficient use of inbound and outbound air traffic."[189] (Previously, the 601st

AOC supported the response to the California wildfires in fall 2007.) The 601st was also able to transfer some combat experience to *Deepwater Horizon.* For example, in *Deepwater Horizon,* the 601st used traditional imagery practices and platforms to find and categorize oil, just as it had done in Iraq when looking for improvised explosive devices.[190]

The staff of the 601st also incorporated important lessons from their Haiti earthquake experience into planning the intelligence support to *Deepwater Horizon.* For example, the priority intelligence requirements and the daily intelligence collection plan, developed with Coast Guard liaison officers, were modeled after the Haiti collection plan.[191]

Even though the staff of the 601st had never used remote-sensing platforms to respond to a spill of national significance, its members were anxious to support the effort. "We don't know anything about oil, but we can sure as hell put one hundred sensors on one thousand targets a day," Stan Gold quoted Colonel Greg Keach, First Air Force J2, as saying during the first week of May.[192] The 601st was the right unit to help alleviate airspace safety issues and support intelligence planning, collection, analysis, production, and dissemination.

Collection Rigor Established

The control of the airspace via the air tasking order improved intelligence collection. Before mid-July, the use and implementation of the ISR planning and execution cycle was not evident in any understood fashion. By directing the 601st AOC to be the sole coordinator of remote sensing and the manager of the priority intelligence requirements (both airborne and satellite), Admiral Allen effectively centralized the remote-sensing effort. Admiral Allen's intent was not only to create unity of command to avoid possible mid-air collisions, but also to improve his awareness of the air picture. He accomplished both.

Another benefit of the ATO was that the 601st AOC conducted the first inventory of all air assets and sensors being used to support the spill response. The IRSCC had started an inventory in early May, but had not completed it. The July 2010 inventory by the 601st AOC was the first comprehensive look at all remote-sensing platforms. Even though the 601st AOC did not have the authority to prioritize information requirements, it did manage the requirements and coordinated remote sensing with the weather, aircraft, and sensor platforms. For example, the 601st AOC requested re-tasking of a Coast Guard C-130J from a lower-priority patrol off the Florida coast to cover the leading edge of the oilspill in the Gulf of Mexico when a NOAA aircraft could not complete that flight because of maintenance.[193] Before the 601st AOC receiving the ATO, it is unlikely another aircraft would have been reassigned to cover the gap. It is important to underscore here, however, that the 601st *asked* the ICP to re-task the C-130J. Though it was positioned in the ICS as the ACC, the 601st AOC did not have the authority to *directly* re-task the aircraft.

The 601st "brought rigor" to the aircraft flight scheduling by cajoling, requesting, and coordinating response flights.[194] As said earlier, even when the daily ATO was in place in mid-July, the 601st was able to directly control only a few flights.[195] Many aircraft legitimately operated in the impacted area for reasons wholly unrelated to *Deepwater Horizon*, such as commercial aircraft ferrying passengers. In any federal response effort, be it a spill of national significance or a Stafford Act response, the Federal Aviation Administration (FAA) maintains authority over the airspace for flight separation. Even though the FAA has authority over the airspace, however, there is no legal mechanism for incorporating non-response aircraft into the ATO or for requiring response aircraft to participate in the ATO. In other words, the only requirement for aircraft operating in and above the response area is to meet FAA regulations. All other aircraft operation is by cooperation.

Responsible Party Issues and Need for Airspace Control

Another issue that makes airspace control for a federal response different from control for a Stafford Act response is the addition of the responsible party. BP executives were becoming increasingly concerned about the disorganized, uncoordinated nature of the aircraft remote-sensing effort, especially considering that they were paying for these flights.[196] These same BP executives argued that better remote-sensing coordination would be more efficient in locating skimmable oil.

Lesson Learned: Airspace Control Is Critical for Collection and Safety

The work of the 601st staff was to control and manage the remote-sensing efforts for the response aircraft, as Admiral Allen and BP wanted, but at the same time the 601st could not direct the compliance of all participating agencies. Despite this, Admiral Allen realized that assigning a Department of Defense command like the 601st, which was practiced in airspace management and remote sensing, was late in coming. In his National Incident Commander report, Admiral Allen recommended that DHS "memorialize in doctrine the use of the Aviation Coordination Center . . . as a matter of course for any national-level response."[197] "If I had to do the whole thing over again, I would've taken the airspace on the first day and set the [601st Air and Space] coordination center up," Admiral Allen later commented.[198]

How It Worked in Practice: 601st Operations in *Deepwater Horizon*

By mid-June 2010, regular intelligence-collection practice was in place to support the response effort. Until that point, Lieutenant Colonel Martinez observed "lots of work, little collaboration, and no staff direction" regarding intelligence.[199] By the third week of June, a daily remote-sensing "battle rhythm" had developed among the 601st AOC, the UAC, and the ICPs. The staff at the 601st developed the collection plan by midday out of informal conversations and discussions with intelligence liaisons from DHS, NGA, USCG, and other participants at the UAC. The Coast Guard liaison to the 601st would approve the collection plan against the requirements, while the commanding officer of the 601st Intelligence & Reconnaissance Division (IRD) approved the collection sensor. Commander Rainey described the relationship with the staff of the 601st IRD as professional and positive.[200] The 601st would then task the assets under its control and have its liaisons at the other ICPs request the coordination of the assets of other agencies not under DoD control.

Satellite Collection Systems: Necessary for SONS Response

Collecting imagery via satellite systems to support a spill of national significance (SONS) response was a new undertaking, but one that ultimately proved helpful. Satellite systems have been used to detect and locate oil on the surface of water, but this was traditionally done for longer-term studies and research. Using satellites to collect imagery to assist tactical oil-skimming operations, however, was a new application that helped primarily to narrow search areas. Multiple commercial satellite systems used a variety of sensors to spot, track, and attempt to identify the oil. The U.S. government also used classified satellite systems to collect imagery about the spill. The satellites proved adept at locating boom, and they were successful in identifying oil, but the sensors were not able to determine if that oil was skimmable.

Several commercial satellite sensors collected imagery in support of the oil-spill response. Dr. Nan Walker of the Louisiana State University Earth Scan Laboratory explained that "there are many, many satellite systems and all of them are a little different" when trying to collect information and intelligence on oilspills.[201] The differences between the satellite platforms are in their frequency of passage over the area to be observed (imaged), the type of satellite, the different companies, and the different sensors carried by the satellite. Frequency of imaging depends on the satellites' orbits, which can vary greatly based on the function of the satellite.

Issues with Commercial and Government Systems: Contracting Concerns

Other important factors in SONS imagery (and disaster response imagery in general) are the specifications of the U.S. government and responsible party (BP, in this case) contracts and arrangements with the satellite companies. These contracts determine with great specificity what, when, and how imagery will be shot, when that imagery will be delivered, and to whom. In addition to the contracts, just interacting with the various GIS companies and countries can be problematic. Each company (and each country) that sells imagery maintains its own processes for data mining and for requesting and obtaining imagery. According to DeWitt Braud, director, academic area, of the Louisiana State University Coastal Studies Institute, some are relatively easy to work with, such as Earth Resources Observation and Science Center, while others, such as the Indian government, typically require a labyrinth of processes in order to obtain imagery.[202]

International Charter on Space and Major Disasters Facilitates Cooperation

Fortunately, an international mechanism exists to streamline imagery collection and processing after a disaster. The International Charter on Space and Major Disasters is an agreement among 16 agencies and countries that have agreed to speed imagery support after a natural or manmade catastrophe. The purpose of the charter is promoting cooperation between space agencies and space system operators in the use of space facilities as a contribution to the management of crises arising from natural or technological disasters, the Charter seeks to . . . supply during periods of crisis . . . data providing a basis for critical information for the anticipation and management of potential crises.[203]

At the request of the U.S. Coast Guard, the U.S. Geological Service (USGS) activated the charter shortly after the *Deepwater Horizon* explosion. This activation streamlined the request and approval process for satellite imagery to support the spill response.

European-, Canadian-, and American-owned commercial satellites took imagery during the *Deepwater Horizon* response. These satellites used radar, measurements of sea temperatures, the reflection of sunlight off the Earth, and high-resolution imagery (photos) to identify and map the oil. The companies discovered that oil "calms" or "settles" ocean waves, permitting synthetic aperture radar (SAR) to detect the presence and extent of the oil.[204] SAR can work through cloud cover, which proved an advantage in hunting the spill. However, SAR created "false positives" when looking at calm water because it indicated that oil might be present there. Tracking temperature anomalies did not prove successful in finding the oil, but high-resolution photo imagery *did* find the oil.[205] The Moderate Resolution Imaging Spectroradiometer (MODIS) sensors on the Terra and Aqua satellites captured photographic imagery of the spill, as shown in Figure 10. MODIS works by capturing the image based on the reflection of the sun's light off the surface of the water, and works best when that reflection is closest to 90 degrees. MODIS, and other photo-imagery sensors, does not work through cloud cover, however, which concerned Rear Admiral Zukunft, as hurricane season moved forward in the Gulf of Mexico.[206]

CAPT Erich M. Telfer

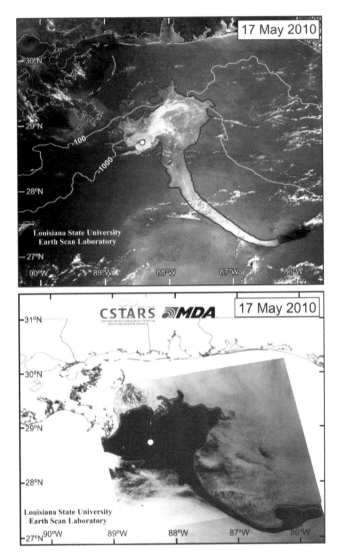

Figure 10: A MODIS image (top) and SAR image (bottom) from May 17, 2010.

Source: Image used with permission and courtesy of Louisiana State University Earth Scan Laboratory.

Commercial satellite imagery was helpful in determining the extent of the spill even though there are myriad sensors and systems available. U.S. government agreements with domestic and international businesses facilitated the collection of satellite imagery, which complemented federal satellite systems.

U.S. Government Sensors

The scope and classification of this paper preclude a deeper technical discussion of the U.S. government noncommercial satellite systems used to support the spill response. Several of these sensors were used to map the oilspill, and NGA and Department of Defense analysts reviewed this data. While the deputy national incident commander, Rear Admiral Neffenger, praised the quality and value of these images, he said that their usefulness was tempered by their classification.[207] The majority of the U.S. government noncommercial images were classified, which meant that all but a few members of the Incident Command Posts and the Unified Area Command could view the images. Other important participants, such as BP, contract, state, and local responders (owing to the lack of a clearance and access), were prohibited from viewing the imagery. U.S. government imagery analysts did produce unclassified text products based on the classified imagery, and these text products contributed to the planning of surface search areas for oil skimming using classified products. But, in the end, the dissemination of the classified imagery was severely limited.

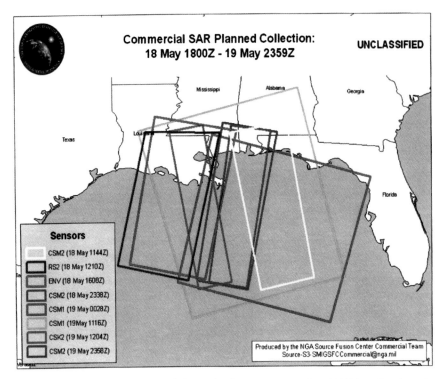

Figure 11: Planned commercial synthetic aperture radar (SAR) collection for May 18–19, 2010.

Source: U.S. Department of Homeland Security, daily brief, "Federal Remote Sensing Situation Report—British Petroleum Oil Spill Response, May 18, 2010," slide 16.

Satellite Imagery Cannot Determine If Oil Is Skimmable

What satellite imagery could not do well was indicate whether oil could be skimmed. This was the case across the contracted commercial satellite collection. The various sensors (radar, photographic, multispectral) could determine the presence of oil, but could not determine if that oil was a light sheen or was thick and goopy. MODIS images could delineate thinner oil from thicker oil, but trained and experienced imagery analysts were needed to review the data.[208]

A deficiency of imagery analysts trained in interpreting MODIS and other imagery of the spill further hindered locating skimmable oil from satellite data.

> **What Is "Skimmable"?**
>
> "Skimmable oil" was a subjective determination. Whether oil could be skimmed was neither a scientific nor a commonly understood definition. The skimmer crews determined if oil was "skimmable" based on their experience and the equipment available on their ship. Thicker oil was more easily skimmed, but wave and wind action reduced the ability to skim oil. Tactical responder Commander Edward "Teddy" St. Pierre, USCG, commented that he observed the crew of some skimmers claim a patch of oil was not skimmable, while the crew of another ship would skim oil of the same thickness.[209]

In the end, satellite imagery was most useful in helping planners refine search areas for others using aircraft and surface vessels. Once the oil was found, based on this data and weather predictions, including sea state and currents, the ICP staff in Houma, Louisiana intelligence liaisons, and staff at the 601st AOC could plan aircraft patrols to support the priority intelligence requirements, skimming operations, and individual agencies' information needs.

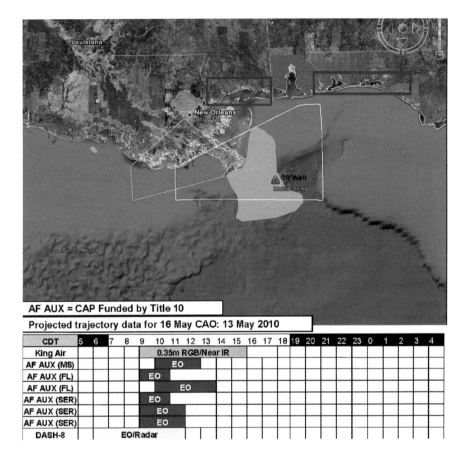

RGB/Near IR – Red, Green, Blue (standard color)/ Near infrared.
EO – electro optical.

Figure 12: The Airborne Remote Sensing Coverage for May 18, 2010, to June 1, 2010. This was part of the actual briefing slide from Federal Remote Sensing Situation Report, also showing the aircraft assets and flight times. These flights were based in part on the satellite imagery from Figure 11.

Source: U.S. Department of Homeland Security, daily brief, "Federal Remote Sensing Situation Report—British Petroleum Oil Spill Response, May 18, 2010," slide 12.

Finding the Oil: Tactical Aircraft and Visual Searches Worked Best

Aircraft used different sensors to find the oil, but visual searches proved the most effective for skimming support. Figure 13 lists a sampling of the aircraft that flew and collected imagery in support of the spill. These planes included Air National Guard RC-26s, U.S. Coast Guard, U.S. Navy, NOAA, Customs and Border Protection, and other federal aircraft. Although not listed in Figure 13, state, local, and BP contract flights also searched for and collected information on oil. In addition to the aircraft scheduled to fly on July 16, a variety of other sensors were also used, including full-motion video, synthetic aperture radar, electro-optical (EO), side-looking aperture radar, multispectral imaging (MSI), and infrared. EO produces black-and-white images, while MSI produces color images. These systems also provided ICP staff and other analysts with additional information, so they could calculate when and where oil might impact marshes, beaches, and other sensitive shoreline areas.

In addition, pilots and flight crews used visual observations (i.e., looking out the window) to find oil. Out of the many sensor platforms used, visual observations offered the best success in supporting tactical skimming operations. Helicopters launched from Coast Guard cutters or shore searched for oil under the direction of the cutters, thus enabling coordinated support for skimming operations.

Tactically, aircraft—including helicopters—were most vital in keeping the Coast Guard cutters and skimmers "on oil."[210] But these visual searches sometimes produced false positives, with aircrews occasionally identifying sargassum seaweed as oil. Searching for and classifying oil is not a core competency of Coast Guard helicopters or crews, and they receive no formal training in this area.[211] In addition, the Coast Guard aircraft lacked real-time video capability to downlink the imagery to surface assets. As explained by Commander Mike Fisher, USCG, the helicopters, plus the RC-26s, flew patrols that searched out and tracked the "leading edge" of the oil approaching the shore.[212] For the tactical responders, the Air National Guard RC-26s were the most useful and successful collection platforms. The RC-26s could provide real-time, downlinked video images to the cutters. These were the only aircraft with this capability.

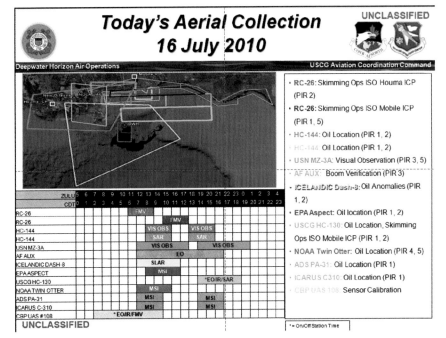

Key: FMV – full motion video

Vis OBS – visual observation

SAR – synthetic aperture radar

EO – electro optical

SLAR – side looking aperture radar

MSI – multi-spectral imagery

IR – infra red

Figure 13: A single day's aerial collection plan, July 16, 2010. In addition to the map graphic, the graph lists which asset would fly, when it would fly in both "Zulu" (Greenwich Mean Time) and Central Daily Time (CDT), and the type of sensor the aircraft possessed. The right box again lists the aircraft and which 601st PIR the flight was collecting against.

Source: Capt Caesar Kellum, U.S. Air Force, 601st Air and Space Operations Center/ ISRD AFNORTH, "Incident Awareness & Assessment (IAA) Support to Deepwater Horizon," briefing presentation, undated but prepared post-summer 2010, slide 8.

Despite the many aircraft and sensors listed in Figure 13, visual observation by pilots and aircrews proved the most successful. One airframe, however, proved its utility above all the others in finding and tracking oil.

Air National Guard Key Asset: RC-26 Aircraft

The Air National Guard contributed to the spill response with their RC-26s. The first RC-26 began working in early June. The Emergency Management Assistance Compact (EMAC) funded and supported the RC-26s. This compact exists between the Air National Guard and the states. Under the EMAC, the locations of the RC-26s among the states change from year to year, but, fortunately, one was stationed in Mississippi at the time of the *Deepwater Horizon* response, and was the first such aircraft on the scene. The success of the RC-26 from Mississippi led to a second plane deployed from West Virginia. Then, a third Air National Guard crew was added, which permitted two 3.5-hour flights per day. The RC-26s gave the tactical responders a tool that they had never used before.

Happy Accident: How the RC-26s Were Requested in the First Place

According to Captain Kellum, the acquisition of the RC-26s was "the single most valuable, important resource request made" in response to the spill.[213] But the initial request came out of a casual conversation during a work break, instead of through careful planning. Two officers in the 601st were taking a short break from the response effort when one of them came up with the idea of real-time streaming video support.[214] The air boss drafted an official Resource Request Message on June 1 following the discussion.

The two RC-26s provided real-time video data to the Coast Guard cutters that were skimming or directing skimming operations. Coast Guard buoy tender Commander Jeff Randall described relying more and more heavily in June and July 2010 on the direct video images provided by the RC-26.[215] Commander St. Pierre said that the RC-26s directly supported skimming operations 85 percent of the time they were flying, while other aircraft only supported the ships 20 percent of the time they flew.[216] Obtaining and using tactical air patrols to support skimming operations was the "most difficult" battle St. Pierre fought.[217] This was the first time that RC-26s had ever been deployed to support U.S. Coast Guard assets and the subordinate vessels of opportunity (VOO).[218] Captain Kellum also praised the work of the RC-26s, commenting that they performed extremely well in supporting tactical skimming.[219]

Figure 14: An April 30, 2010, image taken of the *Deepwater Horizon* spill.

Source: Incident News, "Imagery from 30 Apr 2351 UTC," accessed February 16, 2012, *http://www.incidentnews.gov/entry/526479*.

Other Successful Planes

One other aircraft and sensor used imagery to tactical success. NOAA flew a Twin Otter aircraft with an Ocean Imaging sensor onboard. Ocean Imaging is a company that specializes in GIS and satellite imagery analysis and production. The Twin Otter used an MSI-like spectral imager, which could detect skimmable oil based on its thickness. After the aircraft landed, the crew quickly downloaded the images and e-mailed them to the ICPs, intelligence liaisons, and other support staff. This process worked fairly well. In one case, on July 29, 2010, an image was taken at 10:10 CDT, and it was e-mailed to the ICPs at 12:02 CDT, only two hours later. Also, the Icelandic Coast Guard contributed a Bombardier Dash-8 aircraft to serve as a command and control asset, in addition to spotting oil.[220]

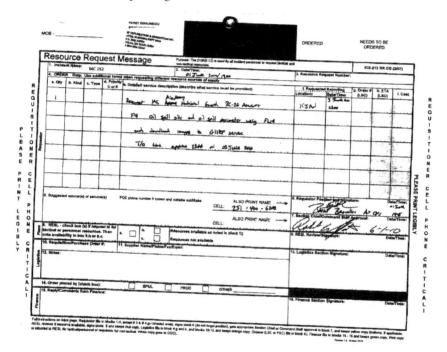

Figure 15: The Resource Request Message dated June 1, 2010, for the RC-26 aircraft. One Intelligence Staff Officer called this "the single most valuable, important resource request made" during *Deepwater Horizon*.

Source: USAF 601st Air and Space Operations Center photo taken by and courtesy of Maj Caesar Kellum, PED Team Chief.

Using Self-Locating Data Marker Buoys

Coast Guard aircraft and cutters attempted to track the oil during darkness by using self-locating data marker buoys (SLDMBs). The Coast Guard usually uses these buoys in search-and-rescue cases. During *Deepwater Horizon*, aircraft crews dropped the buoys out of the aircraft into oil slicks to drift with the wind and current, so that the skimming forces could follow the oil overnight for immediate skimming in the morning.[221] According to Commander Jordan, this method of dropping buoys was "only somewhat effective" because there were not enough of them.[222]

The Coast Guard buoy tenders also used SLDMBs that proved "extremely effective" in relocating skimmable oil and getting surface assets to that oil.[223] This freed aircraft to search other areas because the SLDMBs were "on" the oil. In one day, a Coast Guard cutter recovered over 3,000 barrels (120,000 gallons) of oil because the crew followed an SLDMB that had floated with a slick. Unfortunately, the Coast Guard logistic support centers lacked a sufficient number of these data marker buoys for all of the cutters to use them in tracking oil.[224]

Using People Ashore: Shoreline Cleanup and Assessment Teams

The ICPs also collected information via shoreline cleanup and assessment teams (SCATs). These groups traveled to impacted shore and marsh areas to observe where—and to what extent—oil washed up from the spill. The oil was not easy to find, and these teams had to really hunt for it, often driving an hour or more from the ICPs.[225] The SCATs also performed well in collecting information about boom; specifically, whether it was in the proper location or "off station." The members of the SCATs considered themselves "intelligence officers for the cleanup," and each evening they brought their reports back to ICP Houma, where they passed their observations to a team of analysts.[226]

Collection by Surface Assets (Ships) Proved Least Effective

As mentioned earlier, oil is difficult to observe from the surface, even from the bridge wing of a ship 30 feet above the water. Surface searches by Coast Guard cutters and other skimming vessels without the aid of overhead imagery or aircraft flights proved very difficult.[227] In 1- to 2-foot seas, the crews of the 225-foot Coast Guard buoy tenders had to be as close as 1,000 feet to spot a moderate oil slick consisting of 500–1,500 barrels. Such a slick would be 30–50 feet wide and 1–2 nautical miles long.[228] Mid-morning and mid-afternoon were the best times of the day to spot oil because of the angle of the sun. Spotting oil at night by visual observation was impossible. The

Coast Guard cutter crews did have some success in using shipboard forward-looking infrared to identify oil based on temperature differences, especially in early evening and early morning.[229]

Summary of Collection Challenges and Successes

Intelligence collection, once up and running, went well, but it was initially disorganized and uncoordinated. And after the information was collected to support the *Deepwater Horizon* response, what did the collectors think happened to the information they obtained? Stan Gold was unsure and said, "I don't know. It goes into 'the [intelligence] pool.'"[230] But a "pool" would suggest a common repository for this information where analysts could evaluate it. This was hardly the case.

Step 4: Analysis

While responders collected a considerable volume of information about the spill, the mismanagement of that data confounded analysis. In particular, an absence of sufficient IT system support prevented timely cataloging and production of imagery. Despite this, tactical responders interviewed for this project understood the inherent need for intelligence within their decision-making process and suggested a model for future spill response.

Staffing Issues: Trained Imagery Analysts Required

The ICPs lacked a methodology for using the analyzed imagery even when the data was available.[231] However, even though the ICPs had intelligence liaisons practiced in imagery analysis, those analysts were neither trained nor experienced in exploiting imagery to find skimmable oil. However, they examined the imagery nonetheless. DHS and NGA intelligence representatives were imbedded in the UAC in early May, and in the ICPs by early June. During this timeframe, there was little interaction among the intelligence staffs of the ICPs and the UAC. No one quite knew what to do with the intelligence analysts. In one instance, Lieutenant Colonel Martinez was placed in the operations section as a shift lead, which completely disregarded his experience and value to the remote-sensing effort.[232] Martinez also noticed that members of the operations and planning staff would go for days without reviewing the daily IRSCC situation picture.[233]

Analysis Issues: Deriving Meaning, Need for IT Systems

Deriving meaning from the data turned out to be more difficult than collecting the information. There was no shortage of data. "We were swimming in sensors and drowning in data," commented Captain Kellum.[234] Identifying and classifying oil based on imagery requires skill and practice. "You don't just look at the image and know what you're seeing," DeWitt Braud explained during an interview.[235] Still, NGA and IRSCC analysts impressed the senior response decisionmakers. "NGA was superb," Rear Admiral Neffenger commented, and added that he was "astonished" at how well satellite imagery could find and track boom.[236] Neffenger also relayed a story about an NGA analyst quipping that he did not care if he was searching out foreign tanks or oil; he just liked analyzing and deriving information from imagery. But viewing and understanding the images were not the only challenges to analysis; challenges existed with computer systems supporting the remote-sensing data.

In particular, the intelligence effort was hampered by insufficient hardware and software. The responders lacked computer networks and systems to exploit the remote-sensing imagery. The amount of data proved overwhelming. Neffenger described it as like a library without a card catalog; the responders knew imagery had been captured, but they did not know how to retrieve, view, and manage the imagery.[237] The systems that were being used by BP, NOAA, NGA, the Department of Defense, and the Coast Guard could neither share information nor collaborate. This motivated Stan Gold to comment, "The problem was that [the intelligence support] was being done in a vacuum."[238]

According to GIS professionals Andrew Stephens and Devon Humphrey in their June 9, 2010, open letter posted on the Internet, BP-employed contractors recognized this deficiency, as did the U.S. government responders. Geospatial information systems are essential to oilspill response, to depict and predict the location and the movement of the oil.[239] Unfortunately, as David Gisclair explained to me in an e-mail, "During the first four weeks of the *Deepwater Horizon* spill, the responders lacked the proper geospatial mechanism (system design and operational hardware and software) to acquire, catalog, store, analyze and display the multitude of data streams that decisionmakers needed to execute a well-coordinated response."[240] In one example, NOAA convinced the NGA, BP, and the Coast Guard to use a file transfer protocol that Commander Dash described as "1975 technology."[241]

This protocol served well for academic and strategic oil-flow modeling, but it completely failed to handle the amount of data required to make it useful as a response tool.

Real-Time Oil Locations Not Known

NOAA predicted the flow and future location of the oil. Using aircraft remote-sensing flights and satellite imagery, NOAA scientists accurately predicted the location of the oil in the coming 48 to 72 hours. NOAA staff used an intelligence cycle (planning, collection, analysis, production, and dissemination) in this process, and integrated that process into the ICP staff. Senior staff members at the UAC and ICP Mobile, Alabama, were pleased with this product.[242] Unfortunately, this practice was not real-time collection, analysis, production, and dissemination for response operations, but a forecasting model developed for long-term scientific research. In the end, this NOAA product lacked the detail needed by the U.S. Coast Guard to conduct skimming operations.[243]

The ICPs rarely put real-time information regarding skimmable oil into the hands of the surface responders.[244] Ironically, the intelligence support staff could accurately predict where the oil could be found in the coming days, but not where the skimmable oil was located at that instant. The tactical responders searching out and skimming the oil, along with the contracted vessels, were keenly aware of the lack of intelligence from the ICPs supporting the skimming operations.

After-Action Study Suggests Integrated Model

Tactical responders understood the inherent need for a functioning logic model, including an intelligence cycle, to better accomplish skimming operations. Coast Guard cutter commanding officers spoke to this need in their after-action case study written in December 2010. The study suggests using the surveil, detect, classify, identify, and prosecute (SDCIP) framework for future SONS and for other incidents involving "significant tactical asset employment requirements."[245] The SDCIP is a systematic model, taught by the Coast Guard for operational planning, that fuses remote-sensing information. Figure 16 shows the model the commanding officers included in their after-action case study.

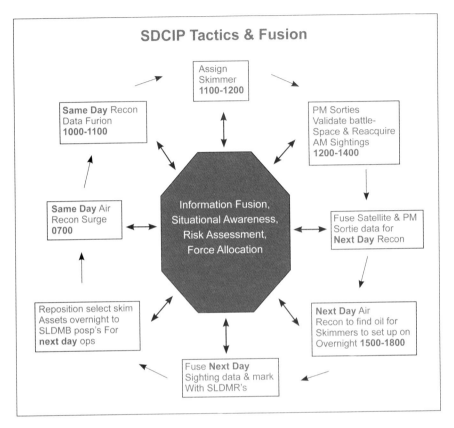

Figure 16: SDCIP Tactics and Fusion Cycle as described in the "Deepwater Horizon Response Surface Operations: A Case Study Prepared by Participating WLB Commanding Officers, page 43.

Source: U.S. Coast Guard, John Kennedy, ed., "Deepwater Horizon Response Surface Operations: A Case Study Prepared by Participating WLB Commanding Officers," expanded after action report, unpublished, December 2010.

The intelligence organizations that responded to the spill lacked the hardware, software, and methodology capable of sharing the collected imagery. Rarely during the response were tactical responders provided real-time intelligence about the location and extent of the oil. This gap was so prevalent that Coast Guard commanding officers developed their own methodology to explain how information and intelligence should be used to support spill response.

Looks Like Oil, I Think

The lack of a protocol for spotting oil and classifying information across the dozens of remote-sensing platforms resulted in misidentification of oil. The Coast Guard, NOAA, and, later, the airborne collectors working for the 601st AOC became practiced in observing and classifying oil from aircraft. But this was not the case with dozens of other agencies and contractors flying over the spill and searching for skimmable oil. In addition, inexperience on the part of some imagery analysts led to false positives in locating skimmable oil. Analysts repeatedly misidentified satellite- and aircraft-collected imagery that they concluded showed skimmable oil. When surface assets arrived in the location, the crews observed seaweed and other surface anomalies where the analysts (and some aviators) reported skimmable oil.[246] In their case study, the commanding officers of the Coast Guard buoy tenders recommended that imagery analysts observe skimming operations from the surface to better compare imagery data with on scene oil sightings. This, the case study recommended, would improve analysis.[247]

BP's Firewalls

BP developed a GIS-based system to manage the remote-sensing information, but controversy followed. In their open letter, Stephens and Humphrey described their work to build a "GIS-based Common Operation Picture capability for ICP Houma."[248] Stephens and Humphrey, collectively, have 40 years of work experience, academic study, and GIS development, according to their résumés, and (according to a source who worked very closely with them during the *Deepwater Horizon* response, but asked to not be identified) they are well known and respected within the GIS community. In the open letter, they explained that their work was completed in the almost record time of about one month, late April through May 28, and provided mapping products for the incident commander, BP, military, ICP/UAC staff, and political appointees in Washington, DC.

According to the open letter, members of BP's IT staff placed this COP behind firewalls, which prevented open and unfettered access to the information, contrary to ICS principles. The authors explained how information and imagery collected by the Louisiana National Guard and U.S. Fish and Wildlife Service were placed directly on the BP server behind the firewalls.

BP subsequently denied access to the information to the agencies responding to the spill, including the Louisiana National Guard and U.S. Fish and Wildlife.[249] Lieutenant Colonel Martinez and another source for this project corroborated the claims about the BP firewalls discussed in the open letter, and further added that BP was reticent to share information and intelligence that its staff and contractors produced.[250] This duplicity in not sharing information and exerting control over information collected by other agencies may have been grounded in BP's litigation concerns, Martinez suggested, but it was not in keeping with the direction and spirit of the National Incident Management System. The deputy incident commander, Rear Admiral Neffenger, complained that getting access to data in a useable fashion during the response was frustrating.[251] It is reasonable to conclude that BP's firewalls contributed to the already uncoordinated, poorly managed remote-sensing practices.

Stephens and Humphrey continued to push for an "open, yet secure" GIS system using the best practices from universities, research, and industry.[252] But it appears they pushed too hard. In late May 2010, BP instructed the subcontracted company that employed Stephens and Humphrey that they were no longer welcome on the project. They had been sacked. Regardless of the emotion, the assertions in the open letter appear sound, and underscore the real intelligence challenges of the U.S. government working alongside a private company, the responsible party, during the response to a spill of national significance, while simultaneously directing that operation and the responsible party. Lieutenant Colonel Martinez commented on the organizational difficulties in working with a private company during disaster response. He opined that the chain of authority between the government and the responsible party must be established early and, while the responsible party participates in the effort, the responsible party should not lead that effort.[253]

"There were so many problems with imagery analysis," DeWitt Braud summed up in a follow-up interview.[254] Included in this complaint were system interface problems among agencies using imagery, and the inability to share images in a timely, consistent manner. What the decisionmakers wanted was a GIS-based common operating picture, such as Stephens and Humphrey had attempted to build, but one that was accessible, unclassified, and available to all responders.

Findings and Recommendations

1. The intelligence support plan recommended at the end of Chapter 3 should use the intelligence cycle to provide decisionmakers with the information outlined in the priority intelligence requirements (PIRs, below).

The intelligence cycle remains a valid method for planning, collecting, analyzing, producing, and disseminating intelligence to strategic, operational, and tactical responders. Although the Incident Command System does not use the intelligence cycle, the Department of Defense intelligence staffs *do practice* the intelligence cycle. Because of this, the intelligence cycle also should be practiced during SONS exercises—especially when remote-sensing and GIS operations are used.

2. The Coast Guard should create priority intelligence requirements for a spill of national significance (SONS) and have them validated. I recommend the following PIRs, which are taken from a July 7, 2010, Federal Remote-Sensing Information Report, as a starting point. (See the appendix for the essential elements of information that give more detail to the PIRs.)

PIR 1: Where is the extent of the oil?

PIR 2: Where are the actionable oil patches?

PIR 3: Where is the boom and what is its status (effectiveness)?

PIR 4: What sensitive areas are threatened or impacted by oil?

PIR 5: Where is the affected shoreline (beach, marsh, estuary)?

PIR 6: What environmental factors will affect the movement of oil (12, 24, 48, 72 hours)?

3. After a SONS, the President should immediately delegate complete control of the airspace to the U.S. Air Force above and around the spill site for flight safety and intelligence collection.

The *Deepwater Horizon* response underscores that the Department of Defense is the sole federal entity with the experience, systems, and trained personnel

CAPT Erich M. Telfer

capable of controlling the airspace above and around a SONS. However, in current practice, other departments and agencies are not bound by DoD air tasking orders. It is therefore incumbent upon the executive branch to direct all of the aircraft (government and private sector) operating in a defined response area to comply with the relevant DoD tasking order. In other words, aircraft operating in the response areas must do so with the permission of and in coordination with the DoD entity controlling all of the flights. The Defense Department (U.S. Air Force), along with the Federal Aviation Administration, would determine the appropriate extent, altitude, and rules governing the airspace to maximize safe flight, response operations, and intelligence collection.

> *4. A remote-sensing concept of operations (CONOPS) should be developed as part of the intelligence support plan that incorporates the federal, state, commercial, and international geospatial information systems that will be used in response to a spill of national significance.*

The remote-sensing CONOPS would define information and intelligence capabilities to respond to the spill. The CONOPS also would establish unity of effort and command among the remote-sensing agencies and delineate the structure of the remote-sensing teams within the UAC and the ICPs. The CONOPS would require enough flexibility to be customized for use in a SONS response, depending on the location and the type of spill.

The CONOPS also should include a remote-sensing methodology based on the intelligence cycle, with particular attention to providing timely, accurate imagery and analysis to tactical responders. The surveil, detect, classify, identify, and prosecute (SDCIP) model developed by the Coast Guard cutter commanding officers combines a decision cycle with a remote-sensing support plan to execute operations. Including such a model in the remote-sensing CONOPS would standardize the practice among the spill responders and set clear expectations for intelligence support to the skimming forces.

Chapter 5
The Common Operating Picture

"Failure of communications appears to be endemic to the human condition."[255]

—Barbara Tuchman

In the aftermath of the *Deepwater Horizon* spill, decisionmakers and staff officers interviewed at the strategic, operational, and tactical levels reported that they needed accurate, timely information and intelligence to respond to a spill of national significance. Specifically, decisionmakers said they needed a common operating picture to answer their questions, and the intelligence staff officers said they needed a method to share their analysis. The Environmental Response Management Application (ERMA) was eventually adopted as a kind of common operating picture after the initial weeks of the spill. ERMA received high marks from strategic-level decisionmakers, but less so from the actual responders. The need for a government-wide common operating picture remains.

Production and Dissemination

Strategic, operational, and tactical decisionmakers first wanted to know what was going on with the spill. They wanted to know the situation. Specifically, the decisionmakers wanted to know where the oil was, where the oil was going, where their ships and planes were, and where the boom was located. Moreover, the strategic-level decisionmakers wanted that information and intelligence to be unclassified and readily available. In short, they wanted a common operating picture. As discussed in Chapter 4, and affirmed by Lieutenant Commander Tabitha Schiro, USCGR, in an interview, there was no shortage of assets collecting images and data, but the timeliness and completeness of those images and information was lacking.[256]

How to best communicate that information to the decisionmakers? Despite the thousands of images taken of the spill to support the response effort, there was no mechanism to organize, display, or manage those images during the first month and a half after the spill.[257] In intelligence parlance, packaging the analyzed information and getting it to the decisionmaker is known as *production and dissemination*. As Captain Kellum put it, "All the remote sensing in the world isn't going to help unless you can get the data to the right people in the right form."[258] That was not done in an efficient manner and, as it turns out, was never fully accomplished.

A common operating picture was the first—and, arguably, most important—need of strategic and operational decisionmakers. Rear Admiral Neffenger, the deputy incident commander, relayed his displeasure that a structure was unavailable to drive information/intelligence management and display this information. He explained how he was frustrated with too many sources of data in addition to unconnected data sources, no commonality in display, and no means of aggregating and sorting that information. According to Neffenger, Admiral Allen said in early May: "I need a platform to display information and I want it to be GIS-based."[259] Neffenger in turn said that he needed to know where the "oil, people, and resources" were. He wanted an application to "clear the fog out and tell me what the picture looks like."[260] The COP had to do more than display information; it needed to contextualize that information and be able to provide meaning.

This need for a common operating picture actually "drove" the National Geospatial-Intelligence Agency coming on board to support the spill response, according to Rear Admiral Neffenger. He and Rear Admiral Zukunft, the second federal on scene commander at the operational level, wanted a single, GIS-based system for tracking the extent of the oil, identifying skimmable oil, and displaying the surface forces, including the thousands of vessels of opportunity. Allen told Neffenger the display must not only serve the Incident Command Post staffs and tactical responders, but must also display the government's response effort. Allen wanted to communicate to government leadership and the public, via the common operating picture, the work being done to fight the spill. Finally, Admiral Allen wanted the common operating picture fully available to the public in real time. This, he argued, would address many of the common questions posed daily, if not hourly, to the leaders of the response and their staffs.

Several GIS-based systems were used throughout the ICPs as common operating pictures. The Department of Homeland Security used a system to manage situational awareness, but it was not used as a common operating picture. BP built a common operating picture, but its limited access prohibited its usefulness to the responders. "It appeared [that] every contractor had their own version" of a GIS-based COP, according to Lieutenant Commander Fisher.[261] And lastly, the National Oceanic and Atmospheric Administration used the Environmental Response Management Application, which by mid-July had developed into the common operating picture Admiral Allen wanted adopted throughout the response effort. According to interviewees, ERMA appears to have offered the best common operating picture during the response, even though it was not widely used until later in the summer and had some shortcomings.

The following sections will review each system in turn and discuss findings and recommendations for a way forward in responding to spills of national significance.

Homeland Security Information System—Capable But with Limited Access

The Homeland Security Information System (HSIN) was designed to serve many purposes, including as a common operating picture. The Department of Homeland Security describes HSIN as "a national secure and trusted web-based portal for information sharing and collaboration between federal, state, local, tribal, territorial, private sector, and international partners engaged in the homeland security mission."[262] HSIN is federally funded and operated by the Office of Operations Coordination and Planning within DHS. HSIN contains information and communication tools, including after-action reports of events, a library of reference documents, web links, event schedules for meetings and exercises, a chat function, mapping tools, and a common operating picture. For the COP, HSIN uses a GIS-based application called GeoPlatform.gov, which represents the response and several categories of data with submenus that the user may turn on to display on the map.

Access to HSIN is limited, however. Although HSIN is unclassified, many of the products placed on the network, including most of the common operating pictures, are marked "For Official Use Only" (FOUO). FOUO is a handling caveat that prohibits dissemination of that material for purposes other than government operations. In other words, FOUO material may not be disseminated to the public or made available via the Internet on an open system.

HSIN is a permission-only network. There are two layers of security a person must pass through before initially gaining access to HSIN. When a person applies for access to HSIN, his or her organization must sponsor that person on a particular "community" or particular "communities" within HSIN. That is, a person cannot gain access to communities within HSIN merely by having access to the site. For example, *Deepwater Horizon* was a "community," as was the Hurricane Irene response and national-level contingency exercises. Within each "community" may be found a COP, documents, information regarding the response, and so forth. This is the first layer of vetting. In addition, the HSIN administrators limit access to those individuals determined to have a need to use the information available on HSIN.

The story of HSIN revealed an important requirement for a common operating picture; in addition to the information needed by decisionmakers, access to such a system played a part. For example, BP built a common operating picture, as discussed in Chapter 4, but it was behind BP's firewalls. Therefore, only BP had access to the information, even though many other organizations and agencies contributed data to the BP COP. Like BP's COP, HSIN was available and widely used during the *Deepwater Horizon* response, but it was behind U.S. government firewalls, preventing industry and other responders from having untrammeled access. The public could gain access to neither of these common operating pictures, despite Admiral Allen's wishes to the contrary.

The Story of ERMA: Selected As the COP

The Environmental Response Management Application (ERMA), developed by the University of New Hampshire (UNH), is a web-based GIS program that pulls and displays data from other systems. In 2006, UNH's Coastal Response Research Center, led by Dr. Nancy Kinner and Dr. Amy Merten, studied how other academic disciplines graphically described and displayed complex environmental systems with multiple data inputs.[263] The researchers wanted to understand *how* the other disciplines captured and showed the data, especially when the information rested in unconnected databases. For example, a tidal region is affected by the wind, air temperature, sea temperature, wave action, moon phase, location, and other geographic factors. The most current information for these vectors may be maintained with different agencies in different systems. Yet to have a current, real-time understanding of the tidal region, all of the information must be culled from the disparate databases and then displayed in a way that is easy to understand. Kinner and Merten understood this, and created the application that became ERMA.

In 2006, the UNH team demonstrated its prototype application to Kent Barton, the NOAA Director of the Office of Response and Restoration, who helped secure approval and funding for further development. In the spring of 2006, the team invited representatives from NOAA Region 1 (in New England), the Environmental Protection Agency, the U.S. Coast Guard, and state representatives of New Hampshire and Maine to view the ERMA prototype. According to Kinner, this "generated a good deal of excitement," and UNH subsequently sought and included user input from these federal and state representatives to guide ERMA development.[264] Federal, state, and local responders then agreed to use ERMA during spill drills in and around Portsmouth, New Hampshire. These practical exercises helped Kinner and Merten improve and refine ERMA by placing real-world constraints and parameters around the use of the application. This garnered further support from NOAA, which Kinner described as "critical" to the growth of ERMA. Once developed, the UNH team turned ERMA over to NOAA to manage.

How ERMA Works

ERMA was designed to cull data from different databases and display it graphically on top of maps and charts. Figure 17 shows the design and architecture of ERMA. ERMA should not, however, be considered a map-making program. The application can store, query, and display spatially referenced data by continually assembling information from other sources.[265] The information ERMA displays is as accurate and current as the information provided by the parent organization in its own database. ERMA can link users to documents, data tables, and live vessel traffic information. In addition, it can incorporate this information onto maritime charts and maps that include weather and navigation data, and create interactive search functions.[266]

The concept and initial research on ERMA appeared promising, but the creators of the application had yet to see if it would work during an actual response.

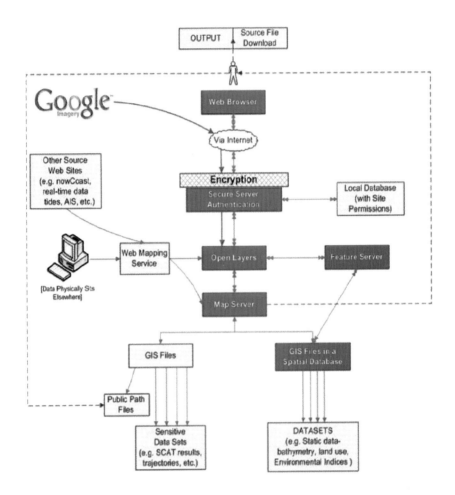

Figure 17: ERMA architecture, as described by the University of New Hampshire's Coastal Response Research Center.

Source: Nancy Kinner, Michele Jacobi, and Amy Merten, "ERMA: Environmental Response Management Application," University of New Hampshire, Coastal Response Research Center, accessed August 15, 2011, *http://www.crrc.unh.edu/erma/ erma_presentation.pdf*, slide number 12.

ERMA: Toward Real-World Applications

ERMA moved quickly from a tested concept to an operational reality. According to Dr. Amy Merten, of the National Oceanographic and Atmospheric Administration, NOAA took ERMA from the UNH development team in 2007 and created a working application for the Environmental Protection Agency in the Caribbean.[267] The Caribbean ERMA has been used to display groundings of small vessels near sensitive areas and for oilspill response drills, and was used during the response to the Haiti earthquake in January 2010. [268] After the earthquake, ERMA integrated imagery of the damage to assist responders and to keep decisionmakers informed of the incident and the international effort to provide assistance. NOAA also created an ERMA site for Puget Sound, Washington, to support a climate-change initiative and to prepare for possible chemical spills in that environment.

Just before the *Deepwater Horizon* explosion and spill, NOAA tested ERMA during the 2010 spill of national significance exercise in March. The exercise scenario envisioned a collision between a large car carrier and an oil tanker in the Gulf of Maine during a severe winter storm, with subfreezing temperatures and reduced visibility.[269] The exercise simulated oil impacting the coastlines of Maine, New Hampshire, and Massachusetts. This was the first time ERMA was exercised on a national level as a situational tool; UNH and NOAA, however, envisioned using ERMA to eventually take on the role of a common operating picture.[270]

NOAA described ERMA during the SONS 2010 exercise as "a web-based open source mapping tool designed to capture and share geographic information used in science *decisionmaking*, both on scene and remotely."[271] (Emphasis added.) And this point bears repeating: NOAA's Office of Response and Restoration envisioned and supported ERMA's development as a *decisionmaking tool*, not merely as a research application. Figure 18 shows the ERMA display during the SONS 2010 exercise and the real-time information available to exercise participants and decisionmakers.

The response to ERMA at the SONS exercise in 2010 was positive. The Shell Oil Company was very interested in using ERMA in response to SONS exercises and in real-world scenarios. Coast Guard planners and responders who took part in the spill exercise argued for adoption of ERMA for broader use

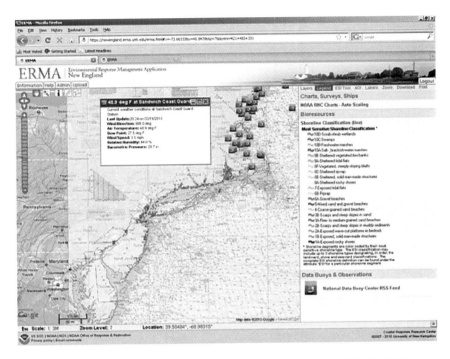

Figure 18: An example of ERMA New England from the SONS 2010 exercise, with NOAA raster charts and buoy locations displayed, in conjunction with the ESI Shoreline Classification layer. The ESI shoreline layer indicates how sensitive the shorelines are to oil. Areas in red are the most sensitive.

Source: National Oceanic and Atmospheric Administration (NOAA), "It's a Drill: SONS 2010," accessed September 6, 2011, *http://response.restoration.noaa.gov/topic_subtopic_entry.php?RECORD_KEY(entry_subtopic_topic)=entry_id,subtopic_id,topic_id&entry_id(entry_subtopic_topic)=808&subtopic_id(entry_subtopic_topic)=2&topic_id(entry_subtopic_topic)=1*

during response operations.[272] The response community would not have long to wait to see ERMA in action.

ERMA Arrives in Response to *Deepwater Horizon*

While ERMA was set up shortly after the *Deepwater Horizon* spill, it took weeks before it served as a common operating picture. Merten and her NOAA staff arrived a week and a half after the explosion of the Macondo well. Sponsored by the NOAA scientific support coordinator and assigned to assist the Coast Guard with situational awareness, some members of the ERMA team set up shop in the Unified Area Command (UAC) at Roberts, Louisiana, with the rest working in the Incident Command Post in Houma. The initial challenge, and a critical component in using ERMA, was determining the information needs of the decisionmakers and understanding their requirements. Merten's team quickly worked to gain consensus from all participants, including BP, to coordinate information needs across the spectrum of responders.[273] This proved challenging because each federal agency, along with the state and local governments and the private-sector responders, was interested in different sets of information to answer questions from the respective chains of command. Complicating this challenge, there was no comprehensive IT system available to manage the volume of data needed.[274] In early June, the staff at the UAC was producing 37 different graphic reports a day to answer data requirements in the different formats wanted by the requesting agencies. This effort consumed considerable staff hours and Stan Gold derided its inefficiency.[275] Decisionmakers responding to the spill needed an easier way to view data and pull meaning from the massive amount of information available.

ERMA Rises to the Challenge

ERMA could help make sense of the massive amount of available data, but it would take a serious discussion between the President and Admiral Allen. That discussion happened almost two months after the explosion, when the response effort appeared to be struggling. Admiral Allen had a "come-to-Jesus talk," as described by the *Washington Post's* Eugene Robinson, with President Obama onboard Air Force One on June 15 about the spill response.[276] The following day, Admiral Allen issued new guidance on the prosecution of the spill response (as described in Chapter 4). At the same time, he also gave straightforward, unambiguous direction to the ERMA team on what he wanted to see.[277] Admiral Allen wanted to know where the surface response ships were, including federal, state, local, and private vessels. He also wanted the vessels of opportunity tracked. The admiral asked that the positions of these vessels be updated every 15 minutes. He wanted to see where the oil was located on the surface and where it was headed. Admiral Allen asked to see the surface and aerial dispersant operations. He wanted all of this data layered on a geospatial information system, and he wanted it on ERMA. In addition to using ERMA as an information and decisionmaking tool, Admiral Allen wanted to use ERMA *as the* briefing tool for visiting VIPs, media, and the public. He wanted the COP to be unclassified and available online to the general public. Finally, the UAC staff had clear requirements and they worked quickly.

ERMA: Problem Solved, But at What Cost?

In less than 24 hours, Merten's team had adjusted ERMA's code to present the majority of the information in the manner asked for by Admiral Allen. Some of the datasets took longer to create, but within a week the team had met all of the admiral's taskings.[278] Had Merten's team known the requirements made by senior decisionmakers sooner, the team could have been providing the desired support all along. And this raises two important points about ERMA, even though *Deepwater Horizon* was just its first real trial as a common operating picture during a crisis: ERMA was flexible and responsive.

ERMA Worked Well for Strategic Decisionmakers

Even though ERMA achieved ultimate success, its usefulness varied depending on the users' place within the response. The more senior a decisionmaker was in the response to *Deepwater Horizon*, the more positively he or she viewed the application. "ERMA was terrific, just terrific," Admiral Allen commented. Rear Admiral Neffenger spoke about how valuable the ERMA team's efforts were in translating data into displayable information.[279] And Rear Admiral Zukunft lauded the diversity of information that NOAA piled onto ERMA for a common operating picture, even including telemetry readings from U.S. Navy unmanned, underwater gliders.[280] As mentioned in Chapter 4, the senior-level decisionmakers wanted to know what was going on at any given moment, and ERMA provided that information to them. ERMA also displayed the response effort being executed against the spill, something Rear Admiral Neffenger underscored during his interview.

The use of ERMA eliminated the need for staffs to provide individual briefs on the status of the spill, which took up a lot of time because decisionmakers, from the White House to state governors, had an almost-constant demand for information about the spill and the response. For example, during the first two months of the response, the UAC produced 37 separate briefs per day to provide situational information. Admiral Allen and Stan Gold described the process of preparing the 37 briefs as cumbersome and time consuming.[281]

But when ERMA matured in the later part of July, the decisionmakers who had previously received the 37 briefs could instead be directed to the ERMA COP, which contained all of the current information. ERMA, therefore, removed the requirement to prepare the many briefs, saving valuable time, effort, and resources.

In addition to reducing work, ERMA helped the strategic-level decisionmakers present information. ERMA excelled as a briefing tool for visiting VIPs. Admiral Allen described what he called a "fiasco" during the crisis, when his staff at the UAC spent "two or three hundred" man-hours preparing a PowerPoint presentation for the DHS secretary just before ERMA came online in July.[282] A very senior official in DHS complicated this effort by becoming intimately involved in the design and production of the presentation to the level of detail of the background colors of the slides. Three days later, Admiral Allen reviewed the ERMA COP with his staff for 15 minutes, and then briefed Vice President Joseph Biden from the COP for 30 minutes. "That's the difference," said Admiral Allen, in the value of using a common operating picture to present information, as opposed to a static brief that displays only one moment in time. ERMA had challenges for other users, though, despite Admiral Allen's praise.

ERMA: Operational and Tactical Responders Less Pleased

Responders working at the operational level in the ICPs and the tactical level both on and over the water found less value in ERMA. "I never found [ERMA] to be a very useful tool," Lieutenant Commander Fisher mentioned during an interview.[283] Part of the problem stemmed from the newness of ERMA and the lack of familiarity with the application. For example, Lieutenant Commander Schiro commented that the ICP spent a great deal of time trying to determine what data ERMA was capturing, what the decisionmakers in the ICP wanted ERMA to capture, and how to extract data (and meaning) from ERMA.[284] This criticism seems one of process as opposed to substance, since the ICP staff received little to no training on using ERMA. A good tool improperly applied in response to a catastrophe will reduce effectiveness.

Another complaint leveled at ERMA was its lack of information about skimmable oil. With a few exceptions, noted in Chapter 4, a paucity of information about skimmable oil was a common issue during the cleanup, regardless of the system used. There existed no information or intelligence system (common operating picture) that provided timely, accurate data on the location of skim-mable oil, including the data in ERMA. This lack of information severely impacted the tactical responders, especially the Coast Guard cutters, which lacked the computers and software to support ERMA.[285] In other words, the Coast Guard cutters conducting the skimming and directing some of the VOOs could not access ERMA. To address this, laptop computers were provided to the cutters to make ERMA available, but the computers met with mixed success.

In particular, the tactical responders were less well-served by ERMA. According to Lieutenant Commander Fisher, the tactical responders became "very frustrated" with the flow of information and intelligence from the ICPs.[286] The remote-sensing images were not easily shared with the tactical responders because the files proved too big for e-mail. Because of bandwidth restrictions, the Coast Guard cutters could not view ERMA either, even after it was adopted as the common operating picture. When the tactical responders provided feedback on problems with ERMA, they saw no change in the application to address their concerns.[287]

Toward a "Common" Common Operating Picture?

Despite the criticisms of ERMA, its use during the *Deepwater Horizon* response proved its worth and sped its exposure and acceptance as a common operating picture.[288] In July 2011, David M. Kennedy, the assistant administrator of NOAA, in his written statement before the Senate Subcommittee on Oceans, Atmosphere, Fisheries, and Coast Guard, said that the ERMA team was named a finalist for the Homeland Security Medal for its assistance to responders reacting to the *Deepwater Horizon* spill.[289] In further response to questions by subcommittee members, Kennedy said ERMA proved so successful in response to the *Deepwater Horizon* spill that he would like to see support and funding for an "Arctic ERMA."[290]

While NOAA is enthusiastic about ERMA, the Department of Homeland Security does not appear as anxious to adopt it. For example, during the week of August 22, 2011, with Hurricane Irene pushing up the U.S. East Coast, DHS asked NOAA to stream the data from ERMA to the DHS system, because it did not want to use ERMA's application to depict the data.[291] It may be that DHS has another common operating picture application in use, such as the Homeland Security Intelligence Network (HSIN). However, I used the

HSIN during a U.S. national-level disaster contingency exercise in May 2011 and found it a slow, cumbersome tool that lacked the comparative refinement and ease of use available with ERMA.

Findings and Recommendations

1. *The Coast Guard should designate a system (hardware and software) to view, analyze, and share remote-sensing imagery during SONS responses among federal, state, local, tribal, and commercial responders.*

One of the most important requirements highlighted by the strategic, operational, and tactical responders to *Deepwater Horizon* was for a GIS-based architecture that can rapidly and accurately display the common operating picture. The vastness of the *Deepwater Horizon* response, coupled with a 24-hour news cycle and a near-insatiable demand for information from the public and senior government officials, required an easily accessible, web-based system to share information in an open manner. Admiral Allen encouraged transparency during the *Deepwater Horizon* response and asked that ERMA be viewable to all.

2. *The common operating picture should be built on an existing system, such as ERMA, that has proven its usefulness during exercises and real-world response operations.*

While other systems are available, ERMA has shown its utility in managing and displaying layered data in a GIS format. The *On Scene Coordinator Report* also recommended that ERMA be adopted as the COP for oilspill response. Using ERMA presents challenges, however, such as the difficulty Coast Guard cutters have in accessing the web page because of bandwidth restrictions for ships that are underway.

3. *The Coast Guard should acquire a system that permits cutters and VOOs to use whatever remote-sensing application the service settles upon.*

Information is the medium of response. Getting the right assets to the right locations when they are needed can be best accomplished through information sharing. This system should be lightweight—and possibly even portable—to permit unclassified Internet use and support of a GIS-based mapping protocol, such as ERMA.

Chapter 6—Conclusion

> "The fact of being reported multiplies the apparent extent of any deplorable development by five to tenfold."[293]

—Barbara Tuchman

This chapter captures the recommendations made throughout this work. Although many of my sources would likely find these recommendations obvious, the epilogue to this work suggests that the organizations responsible for disaster-response planning do not. At the risk of being too obvious, I submit the following.

Intelligence in Disaster Response

The role of intelligence in disaster response is the same as in all other national endeavors: to warn decisionmakers and to help them make better decisions. This was particularly true in the response to the *Deepwater Horizon* explosion and spill of national significance in April 2010. Decisionmakers at all levels, including the strategic, operational, and tactical levels, needed information (and, therefore, intelligence) in a timely, easy-to-use format so that they could deploy personnel and assets to fight the spill.

However, the intelligence support to the *Deepwater Horizon* was uncoordinated and insufficient because of the lack of federal guidance on supporting spill response, a lack of understanding of a worst-case scenario, the absence of a plan for intelligence support (command and control, priority intelligence requirements, intelligence production cycle), and an initial lack of a common operating picture.

On the positive side, strategic decisionmakers praised the efforts of the intelligence staffs in the response effort, and, in the end, despite the handicaps identified in this project, they praised the support that intelligence provided. The support grew steadily throughout the response, and was mature by the time the Macondo well was capped in mid-July.

Standard Organization Required for Intelligence Support to Disaster Response

> *A SONS response, by its very definition of "national significance," should have a separate intelligence section chief who*

CAPT Erich M. Telfer

*reports to the incident commander and supports the other re-
sponse sections.*

Current disaster response guidance, namely the National Response Framework (NRF) and the National Incident Management System (NIMS), do not
adequately address the role and purpose of intelligence in disaster response.
The NRF and the NIMS are purposefully vague to allow, in part, some operational autonomy by the response effort. However, where intelligence support is concerned, there is too much autonomy and too little direction. Even
the location of the intelligence effort within the Incident Command System
is open for interpretation. This slowed the coordination of intelligence support and the ability of intelligence to provide accurate, timely, and "finished"
information, which ultimately hampered unity of effort and command. Intelligence should not be buried within the planning sections as it was in *Deepwater Horizon.*

The Coast Guard Should Have Several Intelligence Support Plans for National Response Efforts

*The Coast Guard should develop several intelligence support
plans that encompass a range of possible maritime events and
responses where the service would be the lead agency. These
plans should clearly delineate intelligence authorities and responsibilities and especially include a remote-sensing concept of
operations.*

Without an understanding of what a worst-case spill could be, the federal
government and the Coast Guard were not prepared to respond to a SONS
of the magnitude of *Deepwater Horizon.* Participants interviewed for this
book repeatedly said the Deepwater Horizon response was new territory and
a never-before-experienced event. Admiral Allen explained that they were
making up the response as they went along.[294] This is hardly encouraging,
and stems from a lack of strategic intelligence and vision. From an intelligence
perspective, it is perfectly conceivable to develop worst-case scenarios for a
range of possible natural or man-induced disasters. The lack of vision extends
to a lack of planning and, specifically, an absence of an intelligence plan to
support a SONS. Several after-action reports from *Deepwater Horizon* have
highlighted the lack of adequate *operational* planning in preparation for the

112

Deepwater Horizon spill (or any spill of that size), but the subject matter expert after-action reports are nearly silent on the need for better *intelligence* planning and integration in spill response.

Training and Education

> *In its intelligence training courses, the Coast Guard should develop and implement a track on intelligence support to disaster response.*

Neither Coast Guard enlisted nor officer intelligence training courses contain material about intelligence support to disaster response. Yet this report and other works have shown the critical intelligence requirements of decisionmakers at all levels after a disaster. These decisionmakers would be better served if the cadre of Coast Guard intelligence officers responding to a national disaster understood and could implement intelligence support to the National Incident Commander and the NIMS after an event.

Airspace Control a Critical Necessity

> *The national incident commander should have complete authority to control the airspace above and around the disaster area.*

In this study, I highlight the use of the Air Force in managing airspace for *Deepwater Horizon*. In reality, lessons from Hurricane Katrina, Deepwater Horizon, the Haiti earthquake response, and the National Level Exercise 2011 all suggest that the President should direct all response agencies using aircraft to fly in coordination with and under the permission of a single authority. Furthermore, that authority should be determined as early in the response as possible, and they should report directly to the national incident commander.

A Common Operating Picture

> *A common operating picture that is accessible to all responders should be developed and agreed upon for use across the disaster response.*

In this study, I recommend building on the Environmental Response and Management Application (ERMA) to develop a common operating picture that includes a web-based GIS capable of layering data to depict the extent of the disaster and the response effort. The COP can be used to develop response actions, inform the public of the progress of the response, and incorporate information and intelligence about the incident. Regarding a maritime response, the COP must be accessible by Coast Guard cutters and other surface response elements. Bandwidth challenges in using web-based applications prevented the Coast Guard cutters from using the ERMA-based COP that was developed during the *Deepwater Horizon* response when the cutters were underway searching for and skimming oil.

Chapter 7
Epilogue

"In this sphere, wisdom which may be defined as the exercise of judgment acting on experience, common sense, and available information, is less operative and more frustrated than it should be."

—*Barbara Tuchman*

Lessons Experienced Are Not Lessons Learned

The Department of Homeland Security and the Coast Guard experienced the intelligence lessons of the *Deepwater Horizon* response, but they did not necessarily learn some of those lessons. This observation is based on my participation in National Level Exercise 2011 (NLE 11) in May 2011. NLE 11 was hosted by DHS and was conducted to gauge a response to a natural catastrophe in the Midwest.

The exercise envisioned two large earthquakes that impacted several states with casualties, significant infrastructure damage, and no communication. The Coast Guard wanted to test situational awareness within the service and DHS. I participated in the exercise, acting as the deputy commandant for intelligence and investigation.

When That Old Man River Shakes

The exercise envisioned two massive earthquakes in the U.S. Midwest that caused thousands of fatalities and injuries, significant power outages, infrastructure damage, and maritime transportation system disruptions in Iowa, Illinois, Missouri, Indiana, Kentucky, Tennessee, and Arkansas. Hundreds of miles of these rivers were closed to commercial traffic and impassable because of damage to the locks and dams. The western river system, including the Mississippi, Ohio, Missouri, and Illinois rivers, is a federal waterway, and during the exercise Janet Napolitano, the DHS secretary, designated the Coast Guard as the lead agency in restoring the maritime transportation system on these rivers.

115

CAPT Erich M. Telfer

A focus of the exercise was to test situational awareness, coordination within the National Response Framework, support to public messaging and warning, and logistical support to a large-scale natural disaster. The most important objective of the exercise, from an intelligence perspective, was to test situational awareness.

Proper situational awareness requires timely, accurate information and, on this count, the Coast Guard and the exercise failed to deliver. First of all, the Coast Guard exercise plan did not build in what some planners called "traditional intelligence play." Instead, intelligence was included and assessed under "situational awareness." But the exercise evaluators (who judge the success or failure of the exercise) wanted to evaluate how well Coast Guard intelligence could support a catastrophic natural event. This was confusing because it focused on only one aspect of intelligence, a common operating picture. To my mind, the lack of broader intelligence play also demonstrated a lack of understanding about intelligence functions in support of disaster response.

The Missing Common Operating Picture

No common operating picture was employed to display and share information among the agencies responding to the exercise. Neither the National Operations Center nor the Coast Guard designated a COP. The National Geospatial-Intelligence Agency used an operating picture, but it was placed on Goggle Earth Pro, which is a proprietary system that includes layered, GIS-based data that can be mined and organized for myriad displays. Unfortunately, if an organization had not purchased the Google Earth Pro application, NGA's operating picture was not accessible.

As mentioned in Chapter 5, the Department of Homeland Security manages a system called the Homeland Security Information Network (HSIN). HSIN is an access-controlled, web-based portal for information sharing among federal, state, local, tribal, territorial, private-sector, and international subscribers. During the exercise NLE 11, the DHS National Operations Center produced a daily COP and posted it on the exercise page. While this operating picture depicted some information regarding the exercise, including updates on casualties, power outages, and earthquake damage, it was not a true COP. Instead, it was a static "snapshot" with limited data as opposed to an active, continually updated common operating picture with layered data.

The lack of a common operating picture was a deficiency that senior Coast Guard leadership had experienced before. Rear Admiral Zukunft, one of the Coast Guard admirals who participated in NLE 11 and the *Deepwater Horizon* response, commented how helpful the ERMA COP, once established, had been in managing and depicting information.[296] Throughout the four-day exercise, the exercise Coast Guard commandant (played by a three-star admiral) commented on the absence of a graphic, geospatial-based product to depict the earthquake impact and the response effort. As previously noted in this work, the Coast Guard Incident Specific Preparedness Review (ISPR) of *Deepwater Horizon* highlighted the concerns of responders who were frustrated with the absence of a single, geospatial-intelligence system–based COP. Ironically, the ISPR argued for a standard, *exercised* COP before an incident occurs.[297]

No Remote-Sensing Coordination

The exercise did not designate an overarching remote-sensing coordination cell (RSCC) to coordinate imagery collection for the exercise. Imagery, from both airborne and satellite collection, would be an obvious tool in assessing the earthquakes' damage, especially considering the communication disruptions throughout the impacted area. Additionally, imagery would have contributed to a common operating picture (had one existed). In the exercise, the Coast Guard National Command Center and the Coast Guard Atlantic Area Command drafted imagery collection requests and routed them to the Coast Guard Intelligence Coordination Center (ICC) in Suitland, Maryland, when no other RSCC could be identified. It was thought that the ICC would be the most effective Coast Guard unit to coordinate the requests. Even though testing remote sensing was not a goal of the exercise, testing the ability to maintain situational awareness was the first objective of the exercise. Imagery would have helped to maintain awareness for strategic, operational, and tactical decisionmakers.

The exercise also failed to designate an authority to coordinate and control aircraft responding to the earthquakes. After the mock earthquakes on May 16, 2011, when the exercise began, federal, state, and local agencies simulated the deployment of hundreds of aircraft into the impacted area. For example, on the first day of the exercise, the Coast Guard simulated flying fixed- and rotary-wing aircraft to staging areas in Arkansas and East St. Louis.

The Coast Guard also flew damage-assessment operations to determine the impact of the earthquake on infrastructure. Unfortunately, these flights were not centrally coordinated by the exercise for information collection, and it appeared the Coast Guard conducted these mock assessments of its own initiative. As with the *Deepwater Horizon* response, flight safety became a concern with so many aircraft operating above the impacted area without coordination. As I recommended in Chapter 4, the exercise response needed an air tasking order or equivalent mechanism to ensure flight safety and to coordinate information collection via remote sensing.

Validated Intelligence Requirements

The exercise did include the important first step of the intelligence cycle, which is the validation of information requirements. This helped to meet the Coast Guard's first objective of exercising situational awareness. By presenting the information requirements to the Coast Guard strategic decisionmakers early in the response, information and intelligence collection could be focused on the needs of the decisionmakers. In a real-world response, the requirements would have driven subordinate coordination and collection, especially regarding the extent of the earthquake damage.

The validation of information requirements during the exercise worked in the following way: the USCG NCC intelligence staff drafted Commander's Critical Information Requirements (CCIRs) within an hour after the exercise earthquakes struck at 1000 EDT. The exercise commandant reviewed and validated the CCIRs at the first meeting of the Coast Guard senior leadership, two hours after the earthquakes. The CCIRs consisted of priority intelligence requirements and friendly forces information requirements. Next, DHS Secretary Napolitano endorsed the CCIRs when she listed her information requirements at the conclusion of the exercise's first DHS senior-leadership group video teleconference. The secretary wanted to know about reestablishing communications in the impacted area and the status of search and rescue.

The following day of the exercise, the Coast Guard operational-level command played by the Coast Guard Atlantic Area in Portsmouth, Virginia, validated its own CCIRs, which were almost identical to those of the exercise commandant and the DHS secretary. The Atlantic area commander increased unity of effort in intelligence collection by reviewing and modeling his CCIRs on those of the exercise commandant.

The following table graphically displays the similarities and differences in intelligence support between Deepwater Horizon and National Level Exercise 2011. A green bullet indicates the intelligence function took place successfully throughout the event. A yellow bullet shows that the intelligence function took place successfully for only a portion of the event, and a red bullet denotes that the intelligence function never happened successfully in response to the event.

Event	Intel Supt. Plan	PIR's ID/ Validated Remote	Remote Sensing Coord.	COP	Air Tasking Order
DWH	●				
NLE 11	●	●	●	●	●

APPENDIX
SONS Priority Intelligence Requirements[298]

PIR 1: Where is the extent of the oil?	
EEI 1.1	Report and identify the forward edge of the oil slick.
EEI 1.2	Report and identify composition of oil in leading edge.
EEI 1.3	Report direction of movement.
EEI 1.4	Identify leading edge of oil plume in the northeast Gulf of Mexico.
EEI 1.5	Report estimated time of landfall.
EEI 1.6	When will the oilspill arrive on the coast?
EEI 1.7	What are the current locations, trajectory, and make-up of oil deposits that could impact the Gulf Coast?
EEI 1.8	Area of Interest coverage for Indications and Warning to detect oil for popup areas off Florida coasts with revisit rate to support operations.

PIR 2: Where are the actionable oil patches?	
EEI 2.1	Report size and location of oil concentrations suitable for skimming operations.
EEI 2.2	Report oil type (sheen or heavy crude).
EEI 2.3	Identify patches of mousse or brown oil larger than 100ft by 200ft in the near-shore zone.
EEI 2.4	Identify patches of mousse or brown oil larger than .5 nautical miles (nm) by .5 nm in the off-shore zone.
EEI 2.5	Report any information containing the depth and composition of the oilspill.
EEI 2.6	Locate and report areas of burnable/dispersible oil.
EEI 2.7	Within 2 hours of collection, report and identify forward edge of oil plumes (not sheen) in the northeast Gulf of Mexico that are of interest (ICP Mobile).

CAPT Erich M. Telfer

PIR3: Where is the boom and what is its status (effectiveness)?	
EEI 3.1	Identify the gaps in boom coverage.
EEI 3.2	Report areas of displaced/beached booms.
EEI 3.3	Report areas of improper boom placement.
EEI 3.4	Report booms being washed ashore.
EEI 3.5	Any significant compromise or loss of barrier emplacements or material requiring more than 24 hours to repair or requiring increased personnel activation?
EEI 3.6	Report degraded booms, to include submerged booms and broken booms.
EEI 3.7	Any boating activity or individuals damaging or stealing booms or barricades.

PIR 4: What sensitive areas are threatened or impacted by oil?	
EEI 4.1	Identify any fisheries, wetlands, and wildlife refuges that are threatened/impacted by oil.
EEI 4.2	Credible reports of oil in or near environmentally sensitive areas.
EEI 4.3	Species, numbers of, and location of dead or injured wildlife resulting from oil spill.
EEI 4.4	Any critical environmental areas or geographic protected areas not protected by mitigation booms.

PIR 5: Where is the affected shoreline (beach, marsh, estuary)?	
EEI 5.1	Identify affected beaches, marshes, and estuaries.
EEI 5.2	Report cumulative length (miles) or shoreline oiled since beginning of release—by region, state, and affected county/parish sector.
EEI 5.3	Report length (miles) of shoreline currently oiled—current total (by region, state, and affected county/parish sector).
EEI 5.4	Report width and area of shoreline impact and encroachment into marshes and coastal estuaries (as available at unclassified level).

PIR 5: Where is the affected shoreline (beach, marsh, estuary)?	
EEI 5.5	Report and locate areas containing large concentrations of tar ball substances and areas where substances are most likely to make landfall.
EEI 5.6	Report degree of shoreline contamination.

PIR 6: What environmental factors will affect the movement of oil (12, 24, 48, 72 hours)?	
EEI 6.1	Formation or movement of a tropical storm into the Caribbean Basin.
EEI 6.2	Changes to tides, winds, or currents that might cause the oil slick to shift toward the coast.
EEI 6.3	Weather forecast of winds from north to south.
EEI 6.4	Observed or predicted off-shore winds greater than 10 knots.
EEI 6.5	Seas of the gulf coast greater than 3 feet.
EEI 6.6	Any changes in the Gulf of Mexico loop current that may produce a new eddy current?

Findings and Recommendations

Findings

Few government or academic works focus on intelligence support to disaster response, and even fewer focus on supporting a spill of national significance. Current federal government after-action reports contain little to no discussion of the intelligence support to Deepwater Horizon.

- The lack of an existing intelligence plan to support a spill of national significance hampered the response effort in Deepwater Horizon.

The intelligence function was not optimally organized during the Deepwater Horizon response because it was placed within the planning sections of the Unified Area Command (UAC) and Incident Command Posts (ICPs) even though intelligence had responsibilities across all interests and departments.

Therefore, the intelligence response to Deepwater Horizon lacked unity of command.

- Strategic, operational, and tactical decisionmakers knew their information requirements, but did not properly communicate those requirements to the intelligence officers.

- A significant amount of data was collected (primarily via remote sensing), but analyzing, producing, and disseminating the subsequent intelligence proved difficult.

- Decisionmakers at the strategic and operational levels were more pleased with the intelligence support provided than the tactical-level decisionmakers were.

The absence of an air tasking order that included all flights operating above and around the spill site led to eight near air-to-air collisions and frustrated intelligence collection.

The initial lack of a geospatial information systems (GIS)–based common operating picture (COP) hampered the sharing and flow of information. Eventually, the Environmental Response Management Application (ERMA) was adopted as the COP, but was never fully embraced by all responders. Therefore, the need for a federal-level COP for use in disaster response remains.

Recommendations

- The Department of Homeland Security (DHS) should direct the Coast Guard to develop intelligence support plans for a spill of national significance (SONS) scenario with the following objectives:

  ○ Ensure that the intelligence section stands apart and supports the other sections (operations, planning, logistics, and administration).

  ○ Clearly detail command relationships within ICS.

  ○ Incorporate a significant GIS/remote-sensing role.

  ○ Ensure that the plan is exercised and reviewed regularly.

- Coast Guard intelligence training and education for enlisted members, officers, and civilians should include a track on the role and function of intelligence support to disaster response.

- The Coast Guard should designate a system (hardware and software) to view, analyze, and share remote-sensing imagery during SONS responses among federal, state, local, tribal, and commercial responders.

- An air tasking order (ATO) should be developed and implemented to manage complete control of the airspace above and around the event. Only aircraft authorized to fly, according to the ATO, will operate in support of the response effort.

- The common operating picture should be built on an existing system, such as ERMA, that has proven its usefulness during exercises and real-world response operations.

NOTES

1 The actual spill amount remains disputed, and 750,000 barrels is considered the upper end of the estimates.

2 National Commission on the BP Deepwater Horizon Oil Spill and Offshore Drilling (hereafter Oil Spill Commission), *Deep Water: The Gulf Oil Disaster and the Future of Offshore Drilling* (Washington, DC: Government Printing Office, 2011), 30. This is the official U.S. government report on the Deepwater Horizon explosion and spill.

3 See 40 C.F.R. § 300.5: "*Spill of national significance* (SONS) means a spill that due to its severity, size, location, actual or potential impact on the public health and welfare or the environment, or the necessary response effort, is so complex that it requires extraordinary coordination of federal, state, local, and responsible party resources to contain and clean up the discharge," *http://cfr.regstoday.com/40cfr300. aspx#40_CFR_300p5.*

4 I follow the standard format of capitalizing the name of a vessel throughout this work. When writing of the drilling rig itself, treated as a vessel by the U.S. Coast Guard, I use "DEEPWATER HORIZON." References to the event and subsequent related activities appear as "*Deepwater Horizon.*"

5 Barbara Tuchman, "In Search of History," in *Practicing History: Selected Essays* (New York: Ballantine Books, 1982), 28. I use excerpts from Tuchman's outstanding work to introduce the chapters. In addition to her excellent study and analysis of history, Tuchman did not shrink from examining the truly difficult aspects of history. I first read Tuchman in graduate school, and her collected essays, titled *Practicing History*, should be required reading for new historical researchers. While *The Guns of August and Stilwell and the American Experience in China* earned her Pulitzer prizes, her work *The March of Folly* speaks more clearly for my research here. In *The March of Folly: From Troy to Vietnam*, Tuchman takes a thoughtfully critical view of what happens when governments and organizations make decisions contrary to their best interest. It should be required reading for all government and military decisionmakers.

6 Bob Cavnar, *Disaster on the Horizon: High Stakes, High Risks, and the Story behind the Deepwater Well Blowout* (White River Junction, VT: Chelsea Green, 2010), 93. It is important to note that estimates on the amount of oil released vary considerably and likely will never be definitively known, according to the author's interview with Admiral Thad Allen, as there was no flow-estimate device on the blowout preventer to measure oil moving through. The amount cited here comes from BP estimates on dispersant calculations and the rate of oil flow released daily.

CAPT Erich M. Telfer

NOTES (continues)

7 The EXXON VALDEZ spill in Alaska was the first SONS and motivated Congress to pass the subsequent Oil Pollution Act.

8 U.S. Congress, House of Representatives, Committee on Homeland Security, *DHS Planning and Response: Preliminary Lessons from Deepwater Horizon*, Testimony of Rear Admiral Peter Neffenger, U.S. Coast Guard, 111th Cong., 2d sess., September 22, 2010, *http://www.gpo.gov/fdsys/pkg/CHRG-111hhrg66030/pdf/CHRG-111hhrg66030.pdf*.

9 U.S. Coast Guard, *On Scene Coordinator Report Deepwater Horizon Oil Spill*, Washington, DC, September 2011, *http://www.uscg.mil/foia/docs/dwh/fosc_dwh_report.pdf, xiii*.

10 Eugene Robinson, "Adm. Thad Allen on What Can Be Learned From the Gulf Oil Spill," *Washington Post*, August 10, 2010, *http://www.washingtonpost.com/wp-dyn/content/article/2010/08/09/AR2010080904869.html*.

11 Peter Neffenger, Rear Admiral, USCG, interview by author, February 10, 2011; interview transcribed, National Defense Intelligence College, Washington, DC. *Senior Decisionmaker*.

12 Of a more humorous note is the occasional actual intelligence officer who fancies himself a Hollywood character from a movie. These people are to be avoided if one can manage it.

13 National Geographic *Explorer*, "Can the Gulf Survive?," September 28, 2010, *http://www.youtube.com/watch?v=6n_RptcxsUs*.

14 U.S. Coast Guard, *U.S. Coast Guard: Intelligence (Publication 2-0)* (Washington, DC: Government Printing Office, May 2010), 1, *http://www.uscg.mil/doctrine/CGPub/CG_Pub_2_0.pdf*.

15 Ibid.

16 Ibid, 2.

17 Intelligence organization is sometimes referred to as "planning." To avoid confusion with the earlier discussion about the lack of an intelligence plan, "organization" will be used.

18 Some describe "intelligence" as information that has been analyzed and had valued added to it. Analysis should help to answer the question *"so what?"* about information. Analysis should not only better explain the information, but be able to explain *why* the information matters within the construct of the situation.

NOTES (continues)

19 Some notional intelligence cycles includes feedback as a final step in the process.

20 USCG senior officer discussion. "Maritime response" is a legacy term. The current term is "prevention," but the functionality is the same. The missions of this portion of the Coast Guard are maritime safety, security, and stewardship. Included in these are environmental protection and responding to oil and chemical spills.

21 A note on the word "intelligence": starting with the Federal Response Plan and continuing on, the various federal disaster plans introduce an interesting, and potentially confusing, dichotomy regarding the word "intelligence" that exists throughout subsequent U.S. government publications on disaster response. On the one hand, "intelligence" refers to information that may assist federal law enforcement in prosecuting criminal cases; that is, information used to build cases, establish warrants, and seek criminal or civil redress. On the other hand, the word is also used to mean an ongoing, iterative process expressed in the intelligence cycle. The latter understanding is the one used throughout this paper.

22 Robert T. Stafford Disaster Relief and Emergency Assistance Act, Public Law 93-288, as amended, 42 U.S.C. §§ 5121–5207, and Related Authorities.

23 Ibid.

24 U.S. Federal Emergency Management Administration, *Federal Response Plan* (April 1992), 16.

25 Ibid., ESF 5.

26 Thad Allen, Admiral, USCG (ret.), interview by author, March 11, 2011; interview transcribed, National Defense Intelligence College, Washington, DC.

27 U.S. Environmental Protection Agency, "National Oil and Hazardous Substances Pollution Contingency Plan (NCP) Overview," *http://www.epa.gov/osweroe1/content/lawsregs/ncpover.htm.*

28 U.S. President, Homeland Security Presidential Directive (HSPD-5), "Management of Domestic Incidents" (February 28, 2003), paragraph 16, *http://www.fas.org/irp/offdocs/nspd/hspd-5.html.*

29 U.S. Department of Homeland Security, *National Response Plan* (December 2004), i, *http://www.scd.hawaii.gov/documents/nrp.pdf.*

30 Ibid., 4.

31 Ibid., 9.

32 Ibid., 25.

NOTES (continues)

33 James Jay Carafano and Richard Weitz, eds., *Mismanaging Mayhem: How Washington Responds to Crisis* (Westport, CT: Praeger Security International, 2007), 229.

34 Homeland Security Presidential Directive (HSPD-5), paragraph 2.

35 Carafano and Weitz, *Mismanaging Mayhem*, 254.

36 Discussion of intelligence within the NRF is built on the *National Response Plan and the Federal Response Plan.* However, intelligence is only sparsely described in the *National Response Framework*, mentioned just 20 times compared with the 153 times in its predecessor, the *National Response Plan.* The *Federal Response Plan* mentions intelligence only 12 times. Frequency, however, does not always indicate breadth and understanding.

37 U.S. Department of Homeland Security, *National Incident Management System* (December 2008), i–ii, *http://www.fema.gov/pdf/emergency/nims/NIMS_core.pdf.*

38 Ibid., 28.

39 Ibid., 59.

40 Ibid., 60.

41 Ibid.

42 U.S. Department of Homeland Security, *National Infrastructure Protection Plan* (2009), 13, *http://www.dhs.gov/xlibrary/assets/NIPP_Plan.pdf.*

43 Ibid., 9.

44 Ibid., 11. The "new" terrorist threat described here is hardly new and certainly not novel. See, also, Andrew C. McCarthy, *Willful Blindness: A Memoir of the Jihad* (New York: Encounter Books, 2008), 176. McCarthy argues that counterterrorism investigations by U.S. federal agencies in the 1980s and 1990s, if not foretelling, certainly forewarned that a 9/11-type incident was in the planning stages. The 9/11 attacks were novel in their magnitude, not in their occurrence, just as *Deepwater Horizon* was novel in its scope and impact but not that it occurred.

45 *National Infrastructure Protection Plan*, 57.

46 Ibid., 9.

47 Domestically, ISR is often referred to as Incident, Awareness, and Assessment (IAA), a term that the United States Northern Command (USNORTHCOM) developed to remove the specter that the government is collecting intelligence against

NOTES (continues)

U.S. citizens. The meaning of IAA, however, is identical to that of ISR, and I use "ISR" here as Major Sovada uses ISR throughout her work.

48 *Remote sensing* is "the science of obtaining information about objects or areas from a distance, typically from aircraft or satellites." *http://oceanservice.noaa.gov/facts/remotesensing.html.*

49 Major Jennifer P. Sovada, "Intelligence, Surveillance, and Reconnaissance Support to Humanitarian Relief Operations within the United States: Where Everyone Is in Charge" (Newport, RI: U.S. Naval War College, 2008), 14, *http://www.dtic.mil/cgi-bin/GetTRDoc?Location=U2&doc=GetTRDoc.pdf&AD=ADA484486.*

50 Ibid., 15.

51 Ibid., 16–17.

52 Lieutenant Commander Joyce Dietrich, "The Eyes of Katrina: A Case Study of Incident Command System (ICS) Intelligence Support during Hurricane Katrina" (Washington, DC: National Defense Intelligence College, 2009), 7. While this work is unclassified, the National Intelligence University (formerly National Defense Intelligence College) maintains all students' thesis work on a classified system if a reader wishes to access Dietrich's work.

53 Ibid., 27.

54 Ibid., 30.

55 Ibid., 51.

56 Ibid., 56.

57 National Commission on the BP *Deepwater Horizon* Oil Spill and Offshore Drilling, *Deep Water: The Gulf Oil Disaster and the Future of Offshore Drilling* (Washington, DC: Government Printing Office, 2011).

58 Raffi Khatchadourian, "The Gulf War," *New Yorker*, March 14, 2011, *http://www.newyorker.com/reporting/2011/03/14/110314fa_fact_khatchadourian.*

59 Peter Elkind, David Whitford, and Doris Burke, "BP: 'An accident waiting to happen,'" *Fortune*, January 24, 2011, *http://features.blogs.fortune.cnn.com/2011/01/24/bp-an-accident-waiting-to-happen/.*

60 Evan Thomas and Daniel Stone, "Black Water Rising," *Newsweek*, June 7, 2010, *http://www.thedailybeast.com/newsweek/2010/05/29/black-water-rising.html.*

61 Khatchadourian, "The Gulf War."

NOTES (continues)

62 Lexington, "The Politics of Disaster," *Economist*, May 6, 2010, *http://www.economist.com/node/16060073.*

63 Other government reports, including those from the Department of Homeland Security, are classified and were unavailable for this paper.

64 Thad Allen, "National Incident Commander's Report: MC252 Deepwater Horizon" (Washington, DC: National Incident Command, October 1, 2010), cover letter.

65 Ibid., 3.

66 Oil Spill Commission, *Deep Water*, vi.

67 Ibid., 265.

68 Ibid., 267.

69 Roger Rufe et al., *BP Deepwater Horizon Oil Spill: Incident Specific Preparedness Review (ISPR)*, Final Report (Washington, DC: Government Printing Office, 2011), 67, *http://www.uscg.mil/foia/docs/DWH/BPDWH.pdf.*

70 Mike Levine, "While Slowing BP Oil Spill, Administration Slowed Flow of Information Too, Claims Coast Guard Report," Fox News (March 28, 2011), *http://www.foxnews.com/politics/2011/03/28/slowing-bp-oil-spill-administration-slowed-flow-information-claims-coast-guard/.*

71 Ibid.

72 Rufe et al., *ISPR*, 52. When Coast Guard publications use the term "operational," the equivalent term in a Department of Defense understanding would be "tactical," if referencing the tactical, operational, and strategic levels of war.

73 Ibid., 53.

74 Ibid.

75 Ibid.

76 Ibid., 54.

77 U.S. Coast Guard, *On Scene Coordinator Report*, 31.

78 Nassim Nicholas Taleb, *The Black Swan: The Impact of the Highly Improbable* (New York: Random House, 2007).

79 Amanda Ripley, *The Unthinkable: Who Survives When Disaster Strikes—and Why* (New York: Crown Publishing Group, 2008).

NOTES (continues)

80 Andrew L. Jenks, *Perils of Progress: Environmental Disasters in the Twentieth Century* (Upper Saddle River, NJ: Prentice Hall, 2011).

81 Tuchman, *Practicing History: Selected Essays*, "When Does History Happen?," 26.

82 Cavnar, *Disaster on the Horizon*, 102.

83 Steve Mufson, "Three Books on the Gulf Oil Spill," *Washington Post* (February 13, 2011), *http://www.washingtonpost.com/wp-dyn/content/article/2011/02/11/AR2011021107065.html.*

84 Cavnar, *Disaster on the Horizon*, 79.

85 Ibid., 97.

86 Stanley Reed and Alison Fitzgerald, *In Too Deep: BP and the Drilling Race That Took It Down* (Hoboken, NJ: John Wiley & Sons, Inc., 2011), 2.

87 William R. Freudenburg and Robert Gramling, *Blowout in the Gulf: The BP Oil Spill Disaster and the Future of Energy in America* (Cambridge, MA: MIT Press, 2011), 1.

88 Ibid., 158.

89 Tuchman, *Practicing History: Selected Essays*, "History by the Ounce," 39.

90 Barbara Tuchman, *The March of Folly: From Troy to Vietnam* (New York: Knopf, 1984), 196.

91 Andrew L. Jenks, phone interview by author, January 25, 2011; interview notes summarized, National Defense Intelligence College, Washington, DC.

92 Joel Achenbach, *A Hole at the Bottom of the Sea: The Race to Kill the BP Oil Gusher* (New York: Simon & Schuster, 2011), 77.

93 Jenks interview.

94 Ibid.

95 Oil Spill Commission, *Deep Water*, 265.

96 Cavnar, *Disaster on the Horizon*, 102.

97 Oil Spill Commission, *Deep Water*, 265.

98 Freudenburg and Gramling, *Blowout in the Gulf*, 158.

99 BP, *Gulf of Mexico Oil Spill Response Plan* (Houston, TX: Response Group, June 30, 2009), 481, 4.

NOTES (continues)

100 Oil Spill Commission, *Deep Water*.

101 Ibid., 266. The MMS is now the Bureau of Ocean Energy Management, Regulation and Enforcement (BOEMRE).

102 DHS employee, interview by author, March 2011; interview notes summarized, National Defense Intelligence College, Washington, DC. *Imagery*.

103 Greg Rainey, Commander, USCG, interview by author, November 15, 2010; interview notes summarized, National Defense Intelligence College, Washington, DC. *Liaison*.

104 Stan Gold, USCG, interview by author, December 15, 2010; interview transcribed, National Defense Intelligence College, Washington, DC. *Liaison*.

105 Ibid.

106 DHS employee interview.

107 Or "Shoreline clean up and assessment technique"; I have seen it designated both ways.

108 I extrapolated this number based on my data collection, interviews, and secondary sources. I think 100 is a conservative number and the actual number of intelligence officers deployed is likely higher.

109 Remso Martinez, Lieutenant Colonel, USA, National Geospatial-Intelligence Agency, interview by author, March 25, 2011; interview notes summarized, National Defense Intelligence College, Washington, DC. Imagery.

110 Gold interview.

111 Martinez interview.

112 Gold interview.

113 Ibid.

114 Allen interview.

115 DHS employee interview.

116 Ibid.

117 Ibid.

118 Martinez interview.

NOTES (continues)

119 Captain Colin Washburn, USAF, 601st AOC, interview by author, February 22, 2011; interview notes summarized, National Defense Intelligence College, Washington, DC. *Imagery.*

120 Neffenger interview.

121 Martinez interview.

122 U.S. Coast Guard, *On Scene Coordinator Report*, 109.

123 FEMA implemented an Intelligence and Investigative Function within the incident command system subsequent to the research and writing of this piece. See *http://www.fema.gov/media-library-data/1382093786350-411d33add2602da-9c867a4fbcc7ff20e/NIMS_Intel_Invest_Function_Guidance_FINAL.pdf.*

124 Tuchman, *March of Folly*, 319.

125 Allen interview.

126 Robert Jensen, Commander, USCG, interview by author, January 4, 2011; interview notes summarized, National Defense Intelligence College, Washington, DC. *Tactical Responder.*

127 Gold interview.

128 Caesar Kellum, Captain, USAF, interview by author, February 22, 2011; interview notes transcribed, National Defense Intelligence College, Washington, DC. *Imagery.*

129 DHS employee interview.

130 Martinez interview.

131 They were Houma, Louisiana; Mobile, Alabama; Florida Peninsula; and Galveston, Texas.

132 Rainey interview.

133 Washburn interview.

134 U.S. Department of Homeland Security, daily brief, "Federal Remote Sensing Situation Report—British Petroleum Oil Spill Response, May 18, 2010," slide 5. This list should not be viewed as the universally agreed-upon requirements, but instead as the requirements the IRSCC staff and other intelligence liaisons and staffs understood to be the information needs of the strategic and operational decisionmakers.

NOTES (continues)

135 Gold interview.

136 Ibid.

137 U.S. Coast Guard, *On Scene Coordinator Report*, 31.

138 Robert Jordan, Commander, USCG, interview by author, January 20, 2011; interview notes summarized, National Defense Intelligence College, Washington, DC. *Tactical Responder*.

139 Achenbach, *A Hole at the Bottom of the Sea*, 184.

140 Neffenger interview.

141 Richard Timme, Commander, USCG, interview by author, January 31, 2011; interview notes summarized, National Defense Intelligence College, Washington, DC. *ICP Staff*.

142 Paul Zukunft, Rear Admiral, USCG, interview by author, February 15, 2011; interview transcribed, National Defense Intelligence College, Washington, DC. *Senior Decisionmaker*.

143 Gold interview.

144 DHS employee interview.

145 Neffenger interview.

146 U.S. Coast Guard, *On Scene Coordinator Report*, 78.

147 Cavnar, *Disaster on the Horizon*, 112–13.

148 U.S. Coast Guard, *On Scene Coordinator Report*, 121.

149 Zukunft interview.

150 Ibid.

151 John Kennedy, ed., "Deepwater Horizon Response Surface Operations: A Case Study Prepared by Participating WLB Commanding Officers," expanded after-action report, unpublished (December 2010), 51.

152 Ibid.

153 Ibid.

154 John Kennedy, Commander, USCG, interview by author, January 4, 2011; interview notes summarized, National Defense Intelligence College, Washington, DC, 7. *Tactical Responder*.

NOTES (continues)

155 U.S. Congress, House of Representatives, Committee on Oversight and Government Reform, Testimony of Governor Haley Barbour of Mississippi, 112th Cong., 1st sess., June 2, 2011, 2, *http://oversight.house.gov/wp-content/uploads/2012/01/Haley_Barbour_Testimony.pdf.*

156 David Gisclair, technical assistance program director, Louisiana Oil Spill Coordinator's Office, e-mail message to the author, October 7, 2011.

157 Chris Lucero, Lieutenant, U.S. Coast Guard, "USCG Remote Sensing Coordinator Statement: Non-BP Signature and Loss of Deepwater Horizon Remote Sensing Concept of Operations Document" (Unified Area Command, Deepwater Horizon MC-252, Memo 16470 to File, August 3, 2010), Enclosure 4. LT Lucero's memo contains the subsequent e-mails detailing the CONOPS to the IRSCC, the legal questions, and Randall's final response that the signed CONOPS had been lost.

158 Chris Lucero, Lieutenant, U.S. Coast Guard, follow-up phone interview, handwritten notes, June 2, 2011. The information in the balance of this paragraph is derived from this phone interview.

159 Deepwater Horizon Unified Area Command, "Remote Sensing CONCEPT OF OPERATIONS: Deepwater Horizon Response" (New Orleans, LA, July 5, 2010), 2.

160 Ibid., 3.

161 Allen interview.

162 Ibid.

163 Gold interview.

164 Ibid.

165 DHS employee interview.

166 Russell Dash, Commander, USCG, interview by author, January 28, 2011; interview notes summarized, National Defense Intelligence College, Washington, DC. *ICP Staff.*

167 Dash interview.

168 Martinez interview.

169 Rainey interview.

170 Dietrich, "The Eyes of Katrina," 51.

NOTES (continues)

171 U.S. Coast Guard, *On Scene Coordinator Report*, 123.

172 Caesar Kellum, Captain, USAF, 601st Air and Space Operations Center, "Incident Awareness and Assessment (IAA) Support to Deepwater Horizon (DWH)," unclassified brief, undated.

173 Dash interview.

174 Neffenger interview.

175 Gold interview.

176 Allen interview.

177 Dash interview.

178 Ibid.

179 DHS employee interview.

180 Achenbach, *A Hole at the Bottom of the Sea*, 172.

181 Ibid.

182 Ibid.

183 Allen interview.

184 Robinson, "Adm. Thad Allen on What Can Be Learned from the Gulf Oil Spill."

185 Allen interview.

186 Ibid.

187 U.S. Air Force, *Intelligence, Surveillance, and Reconnaissance Operations: Air Force Doctrine Document 2-9* (July 2009), vi, *http://www.fas.org/irp/doddir/usaf/afdd2-9.pdf*.

188 Washburn interview.

189 Lieutenant Colonel Susan A. Romano, "Deepwater Horizon Airspace Activity Now Coordinated at 601st AOC," *Air Force Print News Today* (July 13, 2010), *http://www.1af.acc.af.mil/news/story_print.asp?id=123213296*, 1.

190 Gold interview.

191 Ibid.

192 Ibid.

193 Ibid.

NOTES (continues)

194 Ibid.

195 Washburn interview.

196 Rainey interview.

197 Allen, National Incident Commander's Report, 23.

198 Allen interview.

199 Martinez interview.

200 Rainey interview.

201 Nan Walker, director of the Louisiana State University Earth Scan Lab, interview by author, March 1, 2011; interview notes retained by author, Washington, DC. *Academic.*

202 DeWitt Braud, director academic area of the Louisiana State University Coastal Studies Institute, interview by author, March 1, 2011; interview notes retained by author, Washington, DC. *Academic.*

203 International Charter Space and Natural Disasters, *Charter on Cooperation to Achieve the Coordinated Use of Space Facilities in the Event of Natural or Technological Disasters*, Revision 3, April 25, 2000, Article II, *http://www.disasterscharter.org/web/charter/charter#AII.*

204 Braud interview.

205 Ibid.

206 Zukunft interview.

207 Neffenger interview.

208 Walker interview.

209 Edward "Teddy" St. Pierre, Commander, USCG, interview by author, January 24, 2011; interview notes summarized, National Defense Intelligence College, Washington, DC, 2. *Tactical Responder.*

210 Kennedy, ed., "Deepwater Horizon Response Surface Operations," 21.

211 Timme interview.

212 Mike Fisher, Lieutenant Commander, USCG, interview by author, January 12, 2011; interview notes summarized, National Defense Intelligence College, Washington, DC. *ICP Staff.*

213 Kellum interview.

CAPT Erich M. Telfer

NOTES (continues)

214 Ibid.

215 Jeff Randall, Commander, USCG, interview by author, January 11, 2011; interview notes summarized, National Defense Intelligence College, Washington, DC. Tactical Responder.

216 St. Pierre interview.

217 Ibid.

218 Washburn interview.

219 Kellum interview.

220 U.S. Coast Guard, *On Scene Coordinator Report*, 115.

221 Kennedy interview.

222 Jordan interview.

223 Kennedy, ed., "*Deepwater Horizon* Response Surface Operations," 44.

224 Kennedy interview.

225 Khatchadourian, "The Gulf War."

226 Ibid.

227 St. Pierre interview.

228 Kennedy, ed., "*Deepwater Horizon* Response Surface Operations," 20.

229 Ibid.

230 Gold interview.

231 Dash interview.

232 Martinez interview.

233 Ibid.

234 Kellum interview.

235 DeWitt Braud, director academic area of the Louisiana State University Coastal Studies Institute, follow-up interview by author, June 7, 2011; interview notes retained by author, Washington, DC. *Academic*.

236 Neffenger interview.

237 Ibid.

238 Gold interview.

NOTES (continues)

239 Andrew Stephens and Devon Humphrey, "Letter Addressing *Deepwater Horizon* GIS Data Concerns" (open letter published online), posted June 9, 2010. The letter was subsequently taken down from its original site.

240 David Gisclair, e-mail message to the author, June 17, 2011.

241 Dash interview.

242 Timme interview.

243 Gold interview.

244 Ibid.

245 Kennedy, ed., "*Deepwater Horizon* Response Surface Operations," 40.

246 Ibid., 19.

247 Ibid.

248 Stephens and Humphrey, "Letter Addressing Deepwater Horizon GIS Data Concerns."

249 Ibid.

250 Martinez interview, 4. The other source provided this information not for attribution. The BP staff the author contacted for interviews regarding the GIS system firewalls and other intelligence topics related to this project declined to speak on the matters.

251 Neffenger interview.

252 Stephens and Humphrey, "Letter Addressing Deepwater Horizon GIS Data Concerns."

253 Remso Martinez, Lieutenant Colonel, USA, National Geospatial-Intelligence Agency, interview by author, March 23, 2011; typed responses to written questions, National Defense Intelligence College, Washington, DC. *Imagery*.

254 Braud, follow-up interview by author, June 7, 2011.

255 Barbara Tuchman, *March of Folly: From Troy to Vietnam*, 218.

256 Tabitha Schiro, Lieutenant Commander, USCGR, interview by author, January 26, 2011; interview notes summarized, National Defense Intelligence College, Washington, DC. ICP Staff.

257 Neffenger interview.

258 Kellum interview.

NOTES (continues)

259 Neffenger interview.

260 Ibid.

261 Fisher interview.

262 U.S. Department of Homeland Security, "Homeland Security Information Network (HSIN)," *http://www.dhs.gov/files/programs/gc_1156888108137.shtm.*

263 Nancy Kinner, University of New Hampshire, interview by author, July 29, 2011; interview notes transcribed, National Defense Intelligence College, Washington, DC, 1. *Academic.*

264 Ibid.

265 National Oceanic and Atmospheric Administration, Office of Response and Restoration, "Environmental Response Management Application," Information Sheet, July 2011, *http://archive.orr.noaa.gov/book_shelf/1869_ORR-ERMA-07-11.pdf.*

266 University of New Hampshire, Coastal Response Research Center, "ERMA Presentation," *http://www.crrc.unh.edu/erma/erma_presentation.pdf.*

267 Amy Merten, National Oceanographic and Atmospheric Administration, interview by author, August 19, 2011; interview notes summarized, National Defense Intelligence College, Washington, DC. *Academic.*

268 NOAA, "Environmental Response Management Application."

269 National Oceanic and Atmospheric Administration, "It's a Drill: SONS 2010," accessed August 1, 2011, *http://archive.orr.noaa.gov/topic_subtopic_entry.php?RECORD_KEY%28entry_subtopic_topic%29=entry_id,subtopic_id,topic_id&entry_id%28entry_subtopic_topic%29=808&subtopic_id%28entry_subtopic_topic%29=2&topic_id%28entry_subtopic_topic%29=1.*

270 Amy Merten, National Oceanographic and Atmospheric Administration, follow-up interview by author, August 25, 2011; interview notes summarized, National Defense Intelligence College, Washington, DC. *Academic.*

271 NOAA, "It's a Drill: SONS 2010."

272 Merten follow-up interview.

273 Ibid.

274 Dash interview.

275 Gold interview.

NOTES (continues)

276 Robinson, "Adm. Thad Allen On What Can Be Learned from the Gulf Oil Spill."

277 Merten follow-up interview.

278 Ibid.

279 Neffenger interview.

280 Zukunft interview. These U.S. Navy gliders were basically unmanned UAVs that operate underwater.

281 Gold interview.

282 Allen interview.

283 Fisher interview.

284 Schiro interview.

285 U.S. Coast Guard, *On Scene Coordinator Report*, 192.

286 Fisher interview.

287 Ibid.

288 Kinner interview.

289 U.S. Senate, Committee on Commerce, Science, and Transportation, Subcommittee on Oceans, Atmosphere, Fisheries, and Coast Guard, *Looking to the Future: Lessons in Prevention, Response, and Restoration from the Gulf Oil Spill*, 112th Cong., 1st sess., July 20, 2011, 18, *http://www.gpo.gov/fdsys/pkg/CHRG-112shrg72820/pdf/CHRG-112shrg72820.pdf*.

290 Ibid., 29.

291 E-mail to author, not for attribution. The source had firsthand knowledge of this information.

292 U.S. Coast Guard, *On Scene Coordinator Report*, 202.

293 Barbara Tuchman, *A Distant Mirror: The Calamitous 14th Century* (New York: Ballantine Books, 1978), xviii.

294 Allen interview.

295 Barbara Tuchman, *March of Folly: From Troy to Vietnam*, 4.

296 Rear Admiral Paul Zukunft's comment to the author during an NLE 11 brief, USCG Headquarters, Washington, DC, May 16, 2011.

CAPT Erich M. Telfer

NOTES (continues)

297 Rufe et al., *Incident Specific Preparedness Review*, 54.

298 U.S. Department of Homeland Security, "Federal Remote Sensing Situation Report—Deepwater Horizon Response July 7, 2010," presentation, slides 5–7.

BIBLIOGRAPHY

Achenbach, Joel. *A Hole at the Bottom of the Sea: The Race to Kill the BP Oil Gusher*. New York: Simon & Schuster, 2011.

Allen, Thad. "National Incident Commander's Report: MC252 Deepwater Horizon." Final Report. Washington, DC: National Incident Command, October 1, 2010. *http://www.nrt.org/production/NRT/NRTWeb. nsf/AllAttachmentsByTitle/SA-1065NICReport/$File/Binder1.pdf.*

———. "A Strategic Review of the Gulf Oil Spill." Washington, DC: Center for Strategic & International Studies, November 16, 2010. http://csis. org/event/strategic-review-gulf-oil-spill.

Baron, Gerald. "Unending Flow: Case Study on Communications in the Gulf Oil Spill." O'Brien's Response Management, 2010.

Berinato, Scott. "'You Have to Lead from Everywhere': An Interview with Admiral Thad Allen, USCG (Ret.)." *Harvard Business Review*. Accessed October 25, 2010. *http://hbr.org/2010/11/you-have-to-lead-from-everywhere/ar/pr.*

Blew, Robert A. "Synthesis: Intelligence Support for Disaster." *Military Intelligence Professional Bulletin* (July–September 2008). Accessed November 15, 2010. http://findarticles.com/p/articles/mi_m0IBS/is_3_34/ ai_n42852080/?tag=content;col1.

BP. *Gulf of Mexico Oil Spill Response Plan*. Houston, TX: Response Group, 2009.

Carafano, James Jay, and Richard Weitz, eds. *Mismanaging Mayhem: How Washington Responds to Crisis*. Westport, CT: Praeger Security International, 2007.

Cavnar, Bob. *Disaster on the Horizon: High Stakes, High Risks, and the Story behind the Deepwater Well Blowout*. White River Junction, VT: Chelsea Green Publishing, 2010.

Commission on the Intelligence Capabilities of the United States Regarding Weapons of Mass Destruc-

BIBLIOGRAPHY (continues)

tion. *Report to the President of the United States.* Washington, DC: Government Printing Office, March 31, 2005. *http://fas.org/irp/offdocs/wmd_report.pdf.*

Constantine, G. Ted. *Intelligence Support to Humanitarian-Disaster Relief Operations: An Intelligence Monograph.* Central Intelligence Agency, Center for the Study of Intelligence, 1995.

Dietrich, Joyce. "The Eyes of Katrina: A Case Study of ICS Intelligence Support during Hurricane Katrina." Master of Science of Strategic Intelligence thesis, National Defense Intelligence College, 2009.

Erickson, Paul A. *Emergency Response Planning for Corporate and Municipal Managers.* San Diego, CA: Academic Press, 1999.

Freudenburg, William R., and Robert Gramling. *Blowout in the Gulf: The BP Oil Spill Disaster and the Future of Energy in America.* Cambridge, MA: MIT Press, 2011.

Gold, Herbert. *Haiti: Best Nightmare on Earth.* New Brunswick, NJ: Transaction Publishers, 2001.

Jenks, Andrew L. *Perils of Progress: Environmental Disasters in the Twentieth Century.* Upper Saddle River, NJ: Prentice Hall, 2010.

Kennedy, John, ed. "Deepwater Horizon Response Surface Operations: A Case Study Prepared by Participating WLB Commanding Officers." Expanded after-action report. Unpublished. December 2010.

Klare, Michael T. *Resource Wars: The New Landscape of Global Conflict.* New York: Henry Holt and Company, 2001.

Leech, Garry. *Crude Interventions: The United States, Oil and the New World (Dis)Order.* New York: Zed Books, 2006.

Maass, Peter. *Crude World: The Violent Twilight of Oil.* New York: Alfred A. Knopf, 2009.

BIBLIOGRAPHY (continues)

Mayer, Matt A. *Homeland Security and Federalism: Protecting America from Outside the Beltway*. Santa Barbara, CA: Praeger Security International, 2009.

Moore, Melinda, Michael A. Wermuth, Laura Werber Castaneda, Anita Chandra, Darcy Noricks, Adam C. Resnick, Carolyn Chu, and James J. Burks. *Bridging the Gap: Developing a Tool to Support Local Civilian and Military Disaster Preparedness*. Santa Monica, CA: RAND Corporation, 2010.

Morris, John C., Elizabeth D. Morris, and Dale M. Jones. "Reaching for the Philosopher's Stone: Contingent Coordination and the Military's Response to Hurricane Katrina." *Public Administration Review* 67, supp. s1 (December 2007): 94–106.

National Commission on the BP Deepwater Horizon Oil Spill and Offshore Drilling. *Deep Water: The Gulf Oil Disaster and the Future of Offshore Drilling*. Washington, DC: Government Printing Office, 2011.

North American Oceanographic and Atmospheric Administration (NOAA). "Commercial Remote Sensing Satellite Symposium: Key Trends and Challenges in the Global Marketplace," September 13–15, 2006.

Peterson, Marilyn. *Intelligence-Led Policing: The New Intelligence Architecture*. U.S. Department of Justice, Office of Justice Programs. Bureau of Justice Assistance, September 2005. *https://www.ncjrs.gov/pdffiles1/bja/210681.pdf*.

Phelan, Thomas D. *Emergency Management and Tactical Response Operations: Bridging the Gap*. Burlington, MA: Elsevier, Inc., 2008.

Reed, Stanley, and Alison Fitzgerald. *In Too Deep: BP and the Drilling Race That Took It Down*. Hoboken, NJ: John Wiley & Sons, Inc., 2011.

BIBLIOGRAPHY (continues)

Ripley, Amanda. *The Unthinkable: Who Survives When Disaster Strikes—and Why*. New York: Crown Publishing Group, 2008.

Rufe, Roger, et al. *BP Deepwater Horizon Oil Spill: Incident Specific Preparedness Review (ISPR)*. Final Report. Washington, DC: Government Printing Office, 2011. *http://www.uscg.mil/foia/docs/DWH/BPDWH.pdf*.

Sanger, David E. *The Inheritance: The World Obama Confronts and the Challenges to American Power*. New York: Crown Publishing Group, 2009.

Sovada, Jennifer P. "Intelligence, Surveillance, and Reconnaissance Support to Humanitarian Relief Operations within the United States: Where Everyone Is in Charge." Master's thesis, Naval War College, 2008. *http://www.dtic.mil/cgi-bin/GetTRDoc?Location=U2&doc=GetTRDoc.pdf&AD=ADA484486*.

Taleb, Nassim N. *The Black Swan: The Impact of the Highly Improbable*. New York: Random House, 2007.

U.S. Coast Guard. *On Scene Coordinator Report Deepwater Horizon Oil Spill: Submitted to the National Response Team September 2011*. Washington, DC: September 2011. *http://www.uscg.mil/foia/docs/dwh/fosc_dwh_report.pdf*.

U.S. Coast Guard. *U.S. Coast Guard: America's Maritime Guardian (Publication 1)*. Washington, DC. May 2009.

———. *U.S. Coast Guard: Intelligence (Publication 2-0)*. Washington, DC. May 2010. http://www.uscg.mil/doctrine/CGPub/CG_Pub_2_0.pdf.

U.S. Congress. House of Representatives. Committee on Homeland Security. *DHS Planning and Response: Preliminary Lessons from Deepwater Horizon*. Testimony of Rear Admiral Peter Neffenger, USCG Deputy National Incident Commander. 111th Cong., 2d sess., September 22, 2010. *http://www.gpo.gov/fdsys/pkg/CHRG-111hhrg66030/pdf/CHRG-111hhrg66030.pdf*.

BIBLIOGRAPHY (continues)

U.S. Department of Homeland Security. *National Incident Management System*. December 2008.

———. *National Infrastructure Protection Plan. 2009.*

———. *National Response Framework. January 2008.*

———. *National Response Plan. December 2004.*

U.S. Department of Homeland Security; U.S. Coast Guard. *Statement of Admiral Thad Allen, Commandant, on Disaster Preparedness: How the Coast Guard Compares Today to One Year Ago?* Washington, DC. U.S. Congress. Senate Committee on Appropriations, Subcommittee on Homeland Security. September 7, 2006. *http://www.uscg.mil/history/allen/speeches/docs/7Sept2006DisasterPrep.pdf.*

U.S. Federal Emergency Management Administration. *Federal Response Plan*. April 1992.

U.S. President. Homeland Security Council. *National Strategy for Homeland Security.* October 2007. *http://www.dhs.gov/xlibrary/assets/nat_strat_homelandsecurity_2007.pdf.*

———. Homeland Security Presidential Directive–5. "Management of Domestic Incidents." February 28, 2003. *http://www.dhs.gov/sites/default/files/publications/Homeland%20Security%20Presidential%20Directive%205.pdf.*

———. Presidential Decision Directive 39. "U.S. Policy on Counterterrorism." June 21, 1995.

Vitalis, Robert. *America's Kingdom: Mythmaking on the Saudi Oil Frontier*. Stanford, CA: Stanford University Press, 2007.

Youngman, Judith A. "Preparing for Tomorrow's Missions: An Assessment of Strategic Capability in the United States Coast Guard." Unpublished paper, U.S. Coast Guard Academy, 2006.

ABOUT THE AUTHOR

CAPT Erich M. Telfer is a 1991 graduate of the Pennsylvania State University. His Coast Guard assignments include an equal mix of operations and intelligence. Operationally, he has served both afloat and ashore and, most recently, as the commanding officer of the Pacific Tactical Law Enforcement Team, a countersmuggling unit comprising part of the Coast Guard's Deployable Specialized Forces. His intelligence experience includes tours as a tactical and strategic watchstander, three years as the assistant Coast Guard attaché in Mexico City, a tour as the Deputy Director of Coast Guard Counterintelligence, and service as the Executive Assistant to the Assistant Commandant for Intelligence and Criminal Investigations (CG-2). He has earned a Master's of Military Studies from the Marine Corps Command and Staff College and a Master's of Intelligence Studies from American Military University. In 2010, CAPT Telfer was selected to serve as a research fellow at the Center for Strategic Intelligence Research in the National Intelligence University, where he authored this piece, completing his final research in March 2013. Currently, CAPT Telfer commands the Coast Guard Maritime Intelligence Fusion Center Atlantic in Virginia Beach, Virginia. He lives in Virginia Beach with his wife and children.